Leave It in the Ground

Leave It in the Ground

The Politics of Coal and Climate

John C. Berg

PRAEGER®

An Imprint of ABC-CLIO, LLC

Santa Barbara, California • Denver, Colorado

Copyright © 2019 by John C. Berg

Library of Congress Cataloging in Publication Control Number: 2019944343

ISBN: 978-1-4408-3914-6 (print)
     978-1-4408-3915-3 (ebook)

23 22 21 20 19   1 2 3 4 5

This book is also available as an eBook.

Praeger
An Imprint of ABC-CLIO, LLC

ABC-CLIO, LLC
147 Castilian Drive
Santa Barbara, California 93117
www.abc-clio.com

This book is printed on acid-free paper ∞

Manufactured in the United States of America

Contents

Contents

Acknowledgments

This book needed many years to reach completion, and many people had a hand in it. It began when, after decades of working on standard political science topics, I was inspired by the Kyoto Protocol's coming into force in 2005 to join an effort to create an interdisciplinary environmental studies major at my university. The effort was led by Martha Richmond, who became first director of the new program (to whom I am now, but was not then, married); I succeeded her several years later as she moved on to greater responsibilities. I ended up teaching Introduction to Environmental Studies, which took me beyond my background in politics to familiarity with the ethical, humanistic, and scientific issues at stake. Participation in the new Association of Environmental Studies and Sciences further broadened and deepened my understanding.

The next step came when William Crotty invited me to discuss President Obama's environmental record at a one-day conference at Northeastern University and to contribute a chapter to the ensuing book. Evaluation of Obama proved to be a rich vein that could be mined repeatedly, and I was subsequently invited to update this work by Clodagh Harrington and then by Eddie Ashbee and John Dumbrell. The American Politics Group (APG), a specialist subset of the Political Studies Association, provided a forum for very fruitful discussions from which I learned a great deal.

One of my APG papers led Jessica Gribble of Praeger to ask if I would be interested in writing a book on the politics of coal. While I had not intended that to be my next book, it was an opportunity I could not refuse. Jessica has since been tremendously helpful, guiding me through the process of preparing and revising a proposal, displaying exemplary patience and support as I postponed the initial deadline for submission multiple times, and gently letting me know when it was time to finish it for real.

The staffs of the Sawyer Library at Suffolk University and of the Boston Athenaeum have cheerfully complied with multiple requests for interlibrary loans and remote access to materials and, in the case of the latter, offering one of the best sites for writing in the known universe. My department chair, Rachael Cobb, has been unfailing in her support, encouragement, and feedback.

Finally, I return to Martha Richmond. Our transition from professional colleagues to spouses has brought new meaning to my life. Her unflagging support and encouragement, combined with the stellar example of her rigorous approach to her own research, have given me the strength and energy without which this book would never have been completed.

CHAPTER ONE

Introduction

It can be said that coal has fueled the rise of modern civilization. This a bold statement but justified by history. The essence of the Industrial Revolution was to supplement the energy of humans and animals with energy drawn from other sources. Waterpower had long been in use to turn mill wheels, and coal to fuel forges. However, these raw materials were used for energy in the most basic form. It was only with the development of mechanisms to convert one form of motion to another (e.g., rotary to lateral), and later to convert one form of energy to another (e.g., heat to motion, via the generation of steam) that modern industry began. At first, moving water remained the most common energy source, so factories were built along rivers, but with the invention of the steam engine, coal soon became far more important. Coal-fired steam engines drove the machinery in the factories as well as the locomotives and ships that transported raw materials to those factories and the resulting products to the markets. Oil-based engines eventually replaced coal in driving ships—and, for a time, locomotives and electrical motors replaced the steam engines in factories—however, coal continued to dominate the generation of electricity.

Among other things, cheap energy enabled the development of modern science, which has revealed so much about the nature of the world we live in. Ironically, one of those revelations is that burning coal and other carbon-based fuels does a great deal of damage. Some of this damage, the most immediately serious, is due to the presence of impurities in the coal that is burned. With enough effort, these impurities can be removed from coal, either before or after combustion, but in the long run the greatest damage comes from the inevitable result of burning carbon—the creation of a product of that combustion, carbon dioxide. When carbon dioxide is released, it changes the composition of the atmosphere, causing more of the solar energy that falls to the earth to be captured, leading to a phenomenon now known as the "greenhouse effect."

Without the cheap energy from coal, then, we would not know how harmful coal is. We would also not have cured many diseases, lengthened the human life span, sent astronauts to the moon, or built the internet, among many other things. Our challenge is to eliminate the harmful effects of coal without giving up the benefits. While various proposals to do this are being made, my own view is that the only answer is to phase out coal—to leave it in the ground—while developing new energy sources to replace it. This is a huge puzzle to solve. And in order to understand how huge it is, we need to look even closer at the role of coal in developing our modern civilization.

Coal and Civilization

Coal has been burned as fuel for at least three thousand years. It was used in China for smelting copper from the year 1000 BCE and probably used as a fuel much earlier;[1] the Greek scientist Theophrastus described coal and its use sometime around 300 BCE;[2] and coal cinders have been found in Roman ruins in Britain dating to 400 CE.[3] As mentioned earlier, the use of coal accelerated rapidly with the invention of the steam engine. Richard Heinberg reports that coal replaced wood as the world's most important fuel by the end of the 19th century and remained in that position until it was displaced by oil around 1950.[4] Vaclav Smil estimates that global consumption, measured in exajoules (one exajoule is equal to about 280 trillion kilowatt hours), grew from 0.02 in 1800 to 0.65 in 1900, with rapid growth continuing to 141 exajoules in 2008.[5] Large-scale manufacturing and transportation by water and land all developed first with energy from water, wind, and animals, but coal made it possible to increase the size and scale of all three by many multiples. Without it, the Industrial Revolution would have been far less revolutionary.

Beginning early in the 20th century, petroleum began to replace coal for many uses in advanced industrial countries. Most shipping and many railroad lines switched to diesel fuel, automobiles and airplanes were developed from the start to use petroleum products, and oil and natural gas pushed out coal as a means of heating buildings in winter. However, for many other uses, coal was not replaced by other fuels but by electricity, and much of that electricity was produced in coal-fired power plants.

Thus, as Smil's estimates show, the burning of coal remains near its all-time high level.

Coal has brought the world incredible wealth, wealth used to improve human health and comfort for all as well as bringing more individual benefits to some. Coal means energy, and without prodigious consumption of energy, life as we know it would be impossible. At the same time, the burning of coal is devastating for health and the environment. In the short term, a combination of impurities in the coal and less than total efficiency in combustion produce such toxic substances as sulfur dioxide, assorted nitrogen oxides, particulate matter, carbon monoxide, and mercury, all of which are spewed into the air unless they can be filtered out of the exhaust stream. Such filtering is feasible but expensive. (Only as a result of clean air laws has air quality improved considerably.) However, filtering cannot negate the long-term ill effect of the chief product of burning coal and other fossil fuels—carbon dioxide (CO_2). Although CO_2 is a normal component of the atmosphere, and breathing it has no known direct effects on human health, the increased concentration of CO_2 reduces the capacity of the earth to emit excess heat into outer space, and so gradually the temperature on our planet has increased and the energy of the atmospheric/oceanic system has produced the phenomenon known as "climate change" or "global warming."

An Overview of Coal and Climate

The laws of probability suggest that it is unlikely that humans are the only species in the universe capable of collective learning, which is "the ability to share ideas so efficiently that the information learned by individuals begins to accumulate in the collective memory from generation to generation."[6] Adherents of the Big History discipline propose that all such species must pass through three stages:

> In Stage 1, childhood, these species accumulate a growing body of knowledge about their environment. This gives them increasing power to extract resources from their environment and support ever larger and more complex communities. Barring extreme events such as asteroid strikes, they eventually reach Stage 2, adolescence. In this stage, they have accumulated so much power over their environment that they can now transform their planet, although it is not yet clear if they have the wisdom needed to use their power well. This potential mismatch of power and wisdom may create a bottleneck, difficult to pass through, and this may explain why we have not heard from other such species although we have been listening for over half a century.[7]

Stage three, which could be called "maturity," is the attainment of a sustainable society. A similar point about the destructive power of humans was made, in a theological rather than a cosmological context, by George Perkins Marsh:

> The fact that, of all organic beings, man alone is to be regarded as essentially a destructive power, and that he wields energies to resist which Nature—that nature whom all material life and all inorganic substance obey—is wholly impotent, tends to prove that, though living in physical nature, he is not of her, that he is of more exalted parentage, and belongs to a higher order of existences, than those which are born of her womb and live in blind submission to her dictates.
> . . . While the sacrifice of life by the lower animals is limited by the cravings of appetite, he unsparingly persecutes, even to extirpation, thousands of organic forms which he cannot consume.[8]

It was the development of coal, and later of oil, as sources of usable energy that gave humanity the power of unlimited destruction, moving us into the adolescent stage. The abundant energy fossil fuels provide has made modern civilization, based on scientific knowledge, possible. Without them, we would not have been able to detect the potential catastrophe that these same fuels are bringing upon us. Without burning coal, we would not know that we are now in the bottleneck; unless we stop burning coal, we will not get through it.[9]

Because it is so central to modernity, coal has been close to the heart of climate politics. Ending the combustion of coal would be the single biggest step toward reducing greenhouse gas emissions, but since such a cessation would cause great financial losses for those who own the coal reserves and mines, the proposal to stop burning coal generates a lot of conflict. This conflict is only the latest chapter in the politics of coal. Previous conflicts have involved workers and unionization, the destructive effects of strip mining, and the role of coal in causing acid rain. In each of these areas, the coal mine owners have confronted a progressive coalition over some of the central issues of the day.

The Plan of This Book

The next two chapters of this book will discuss some of the past political controversies involving coal, with particular attention to labor-management conflict, pollution from mines, and air pollution from the burning of coal. We will also take an in-depth look at the relationship between carbon dioxide and the earth's climate and evaluate the possibilities of the various proposals for limiting climate change, which can be divided into two groups: those that allow for the continued burning of coal while its impact on the climate is controlled in other ways, and those that foresee development of alternative energy sources that do not have a negative effect on the climate. The last section of this book, then, examines current and proposed mechanisms for reducing the burning of coal, including international agreements, domestic regulation, and the impact of grassroots climate activism. These three arenas include the United Nations Framework Convention on Climate Change, U.S. government action through legislation and regulation, and direct citizen action by means ranging from climate-sensitive consumption to nonviolent (and occasionally violent) civil disobedience. The book concludes with an assessment of where we are today and a consideration of the prospects for holding the increase in mean global temperature below two degrees Celsius.

The Political History of Coal

The early history of American coal production was marked by violent lawless conflict. Coal mining took miners far beneath the surface of the earth, where they faced the imminent dangers of fire, toxic gases, and cave-ins, as well as the long-term threat to their health from the poisonous environment in which they worked. Miners' attempts to win basic health and safety protections, as well as more adequate wages, were met with strong resistance from the mine owners and operators. The latter were often able to use prosecutors and courts as their allies; however, they did not rely on legal action alone but organized their own private armed forces, often under the auspices of the Pinkerton Agency.[10] After decades of violent and nonviolent struggle, strong labor unions arose in the coalfields and became the foundation of industrial unionism in the United States. While unions and operators sometimes collaborated to seek subsidies for the industry, conflict continued to be a frequent occurrence.

Then, beginning in the middle of the 20th century, technology was developed to mine coal by stripping away the surface material, rather than digging into it. Strip-mining lessened the need for labor, while vastly increasing the negative impact of coal mines on local communities and the environment. Here, too, an atmosphere of semilawlessness prevailed. Many Appalachian rural communities were inhabited by people who lacked clear title to the land they lived on, and mine operators seized the opportunity to obtain land titles through a combination of shady legal maneuvering and brute force. In return, statewide grassroots campaigns sought to ban strip mining. While these campaigns did not succeed any-where, they laid the foundation for national regulation of strip mining.

By the end of the 20th century, new political issues had developed. However, the industry's experience of defying government and taking the law into its own hands continued, increasing the difficulty of government action on such issues. These attitudes were to play an important part in the debates of the 1960s and 1970s around federal air pollution control.

Coal and the Clean Air Act

The killer fogs of London; the air pollution disaster in Donora, Penn-sylvania; and the prevalence of the pollution that became known as "smog" in Los Angeles inspired attempts to protect the air we breathe from pollutants of all types, from radioactive fallout to chemical toxins, with particular emphasis on stationary and mobile sources of pollution: industrial plants and motor vehicles. The struggle for a Clean Air Act became part of the broader environmental movement. This movement constituted a new kind of politics, with many activists who were motiv-ated by general principles rather than their own self-interest. While every-one certainly has a personal interest in cleaner and healthier air, the groups that campaigned for an air pollution law came to be seen as a par-ticular interest group, "the environmentalists." The Clean Air Act created a new regulatory framework, and President Nixon, using his reorganiza-tion powers, created a new agency to administer this and other environ-mental laws, the Environmental Protection Agency (EPA).[11] The EPA was designed to assess existing science to determine when pollutants had a negative impact on human health, and to develop and implement stan-dards to reduce such impacts to acceptable levels. Over the decades, the nation's air became cleaner, but only with considerable controversy. More-over, where coal was concerned, much of the cleanup was achieved by developing "scrubbers" to remove pollutants from the exhaust stream and by opening up the vast coalfields of the Mountain West, where the coal held smaller proportions of sulfur. Coal use did not decline, but the exhaust became cleaner. However, in the 1980s, awareness began to spread of a new problem with coal—its contribution to climate change.

Coal and Climate Change

The remainder of the book focuses on coal as a driver of climate change and various strategies to limit this effect while still meeting society's energy needs. Scientists have known for over a century that carbon dioxide is what is called a "greenhouse gas," so called because it has optical properties similar to those of the walls of a greenhouse, letting visible light through but blocking light in the infrared portion of the spectrum. These properties give the atmosphere a heat-trapping effect: visible light passes through and falls on the surface of land or water, thereby heating them, but the resulting infrared heat radiation is blocked.[12] The higher the concentration of carbon dioxide and other greenhouse gases in the atmosphere, the more heat is trapped. Additional heat means higher temperature, as well as additional energy to cause more frequent and stronger storms.

All that has been known in theory for a long time. However, it was only toward the end of the 20th century that scientists began to notice how fast temperature change was occurring, a realization brought to public notice by the dramatic testimony to a Senate committee by NASA scientist James Hansen in 1988 and then by environmental journalist Bill McKibben's best-selling book *The End of Nature* the next year.[13] Chapters 5 and 6 will present a basic scientific understanding of the greenhouse effect and coal's role in it, and assess the feasibility of burning coal while keeping carbon dioxide out of the atmosphere, as well as the availability of getting the energy we need from other sources. The book concludes with two chapters about the contemporary struggle over the politics of coal and climate.

CHAPTER TWO

The Political History of Coal

The political posture of the coal industry has been shaped by past battles over the harmful consequences of mining and using coal. These include battles with miners over pay and working conditions, battles over the destruction of the landscape by coal mining, and battles over the toxic substances emitted when coal is burned. A brief look at these three areas will help us understand the current battle over coal's impact on the global climate.

Coal and Labor

From the struggles of the fictional miners in Zola's *Germinal*[1] to the defeat of Arthur Scargill's strike against Thatcherism, by way of John L. Lewis's role in founding the Congress of Industrial Organizations (CIO), coal miners have often been in the vanguard of organized labor. While their prominence can sometimes be attributed to the strategic choices made by miners' leaders, it also reflects the centrality of coal to industrialization. For close to a century, coal was essential for powering steamships and railroads, and for the making of steel.[2] The importance of coal as a resource gave the miners added leverage, while simultaneously giving mine owners increased incentive to seek to hold wages down.

Coal Miners in the 19th Century

Prior to the Civil War, most coal operations were small, with only a few workers mining coal that was close to the surface. Miners were treated as subcontractors, working at their own pace and being paid by the ton of coal produced. During the war the needs of both sides for fuel caused the industry to become more tightly organized, a trend that accelerated after the war's end. Railroad companies began to purchase large tracts of coal land and build lines directly to them. Technological improvements and capital investment were required in order to mine coal that was deeper in the ground, particularly if it was below the water table. Mines grew in size, and owners started to hire managers to regulate the pace of work. However, miners continued to be paid by the ton, rather than the hour. As competition in the industry drove the piece rate down, miners reacted in a series of small, mostly unsuccessful strikes.[3]

In September 1869, a fire at a Pennsylvania mine killed 110 miners. The owners of the mine had refused to construct a second exit, or install other safety equipment, and miners had been trapped inside. John Siney, leader of the Workingmen's Benevolent Association (WBA), an industrial union formed a year earlier, arrived on the scene and made a rousing speech, calling on miners to join the union. Many did, and the WBA became the first U.S. miners' union of more than local scope.[4]

Membership in the WBA grew to over thirty thousand members, 85 percent of the anthracite miners of Pennsylvania. Meanwhile, the owners, too, had organized, forming the Anthracite Board of Trade, led by Frank Gowen of the Philadelphia and Reading Railroad and the closely linked Philadelphia and Reading Coal and Iron Company. Gowen wanted to break the union and persuaded the other owners to implement a 20 percent pay cut in the industry, thereby provoking a strike. The strike lasted six months, with violence and death on both sides, until the union was broken and the miners accepted the pay cut and returned to work. Violence continued, however, with extrajudicial killings both by the Pinkerton Agency, which had been hired by Gowen, and by miners allegedly operating through a secret society calling itself the Molly Maguires. Eventually twenty miners were arrested, tried, convicted, and hanged for killings attributed to the Mollies, although the fairness of the trials was questionable. Understandably, it has been difficult to determine the reality of a secret society that kept no records, and controversy over the existence of the Mollies, as well as their alleged links to the Ancient Order of Hibernians (an Irish-American fraternal group) and the WBA continues to this day.[5] The defeat of the WBA ended national unionization of coal mining from 1875 until the formation of the United Mine Workers of America (UMW).

The Rise of the United Mine Workers

The UMW was formed in 1890. Following a wave of strikes that began in 1897, UMW membership grew, reaching a quarter million by 1903.[6]

The UMW sought to organize coal miners in the West as well. It organized a strike in Gallup, New Mexico, in 1903, and soon after won a contract in the mines of northern Colorado. In 1913, it led a strike against a group of mining companies in southern Colorado, including the Colorado Fuel and Iron Company, which was owned primarily by John D. Rockefeller. The strike had seven demands, most prominently an eight-hour day, a 10 percent increase in pay, and company adherence to the existing health and safety regulations. The company refused and evicted striking miners from the company-owned camps where they and their families lived. In response, the miners created tent cities.

On April 20, 1914, a combination of the Colorado state militia and the Colorado Fuel and Iron Company's guards attacked the tent city at Ludlow, Colorado, with machine gun and rifle fire, touching off a pitched battle in which twenty-five people, including two women and eleven children, were killed, an event that became known as the Ludlow Massacre. The miners retaliated with attacks on company facilities and guards over a range of forty miles, but they were ultimately defeated. Estimates of total fatalities ranged from 69 to 199. Although the UMW lost the strike, the event produced a wave of public sympathy for the miners and their cause as well as a sense that company lawlessness needed to be brought under control.[7]

The UMW had been formed as a craft union and continued to function as one during its first decades, affiliating with the American Federation of Labor (AFL). However, labor-management relations continued to be marked by violence, and John L. Lewis, who had become national president of the union in 1911, eventually developed a new vision. Rather than basing itself on a particular craft, the UMW would seek to unite everyone who worked in the coal industry in whatever capacity (except management) as an industrial union. As the Great Depression developed, Lewis sought to spread this vision beyond the miners, at first under the auspices of the AFL's Committee on Industrial Organization, which Lewis had proposed to the AFL in 1934. By 1935, he had grown impatient with resistance to the industrial union concept from traditional craft unionists, particularly William Hutcheson, then president of the AFL. During the 1935 AFL convention, an argument broke out between the two men, culminating when Lewis punched Hutcheson in the nose, knocking him down and thereby helping dramatize the simultaneous exit and expulsion of the industrial unions, which then formed the Congress of Industrial Organizations (CIO) as a competing national federation.[8]

The CIO grew rapidly, aided on the one hand by a wave of militant sit-down strikes during the 1930s and on the other by the passage of the Wagner Act, which provided a process for legal recognition of unions as bargaining agents of the employees of a given company that made both organizing easier and enabled unions to tell those whom it sought to organize that the president of the United States wanted them to join a labor union. Lewis was the leader of this organizing campaign; however, he eventually broke his alliance with the Roosevelt administration and the other unions of the CIO, took the UMW out of the federation, and led a number of strikes during World War II that were widely condemned by the president and by other labor leaders.

More important for the issue at hand, coal began to lose its position as the most important energy source. Concerns with air pollution drove a massive switch from coal to other fuels for home heating, and the postwar rise of the auto industry moved auto manufacturing and oil, rather than railroads and coal, into the centers of both the U.S. economy and the labor movement. Although the U.S. auto industry has now suffered its own decline, and former UMW president Richard Trumka is the current head of the reunited AFL-CIO, Trumka won the election more because of his commitment to a more militant strategy than to the strategic importance of mining.[9]

One result of this history of labor conflict is that most progressives feel a basic solidarity with the miners' union. Such songs as "Which Side Are You On?" and "A Miner's Life" are embedded deeply in left culture. Moreover, progressives of all sorts and labor activists have a long history of working together. On the other hand, the UMW has, not surprisingly, been concerned with the continued viability of coal mining. When the Kyoto Protocol was signed, Trumka, at that time still president of the UMW, opposed it, calling for increased use of coal. Today, as president of the AFL-CIO, Trumka has taken a more progressive line on climate change, declaring, "The carbon emissions from that coal, and from oil and natural gas, and agriculture and so much other human activity—causes global warming, and we have to act to cut those emissions, and act now."[10]

Labor unions everywhere are seeking ways to overcome the "jobs vs. the environment" opposition.[11] Perhaps because of their history of violent confrontation with management, the UMW has shown less tendency than some of the building trades to form a bloc with management against progressive climate policy. However, they do face a real dilemma that involves more than climate policy, but rather extends to other environmental problems of coal mining as well, and they have not yet found an answer to that dilemma.

Strip Mining and Erosion

Although surface mining has been practiced since at least the 16th century,[12] it was the development of heavy machinery in the 20th century that made large-scale strip mining of coal possible. Excavation machines could move up to twelve thousand cubic meters of earth per hour to strip the overburden of soil and rock that covered a seam of coal. After coal was extracted, the overburden from the next strip was pushed into the trench, with the process repeated until the far side of the coal seam was reached.[13] The very large capital investment in the excavation machines and other equipment made it possible for the mining companies to produce coal with a much smaller labor force, a change that was generally good for the investors but detrimental to the communities the mining workforce came from. Strip mining also has detrimental environmental effects, including the immediate destruction of arable land above the coal seams, pollution of rivers by runoff from uncovered minerals and slag heaps, and ecological changes as the mined land was either left in its destroyed condition or rehabilitated to conditions different from the original ones.

Strip mining in Appalachia, where it first developed on a large scale, was quickly perceived as destructive by some local residents, but these opponents found themselves almost powerless to stop its spread.

The Development of Strip Mining

European settlement of the Appalachian region took place prior to enactment of the Homestead Act, and much of it prior to the American Revolution. It was therefore more difficult for settlers in Appalachia to obtain title to family-sized farms than it was for settlers in the Midwest. By the end of the 18th century, much of the land in the region was held by absentee speculators. This situation was reversed by 1880, with a large proportion of the land in family-held farms. However, this change was brought about in good part by squatting, which was simply occupying a piece of land until the farmer's title was recognized de facto.[14] As a result, the title of small farmers to their land was not always firmly established; when extractive industries (e.g., timber as well as coal) began to grow during the late 19th century, experienced land agents could often undermine the title of farmers who refused to sell. Montrie reports that by 1910 large percentages of timber, coal seams, and surface land were controlled by "outlanders."[15]

Some farmers sold mineral rights with the expectation that they would continue to farm their land while coal mining was carried out underneath it. Strip mining changed all that; under the commonly used *broad form deed*, the right to extract coal turned out to include the right to remove the overburden and dump it elsewhere on the property, and the right to erect roads and buildings.[16] Montrie quotes a 1965 letter to the editor of the *Hazard Herald* in Kentucky by activist farmer Clarence Williams:

About two years ago . . . [I] sold right of way to [a] strip mining company. . . . Now it makes me sick to look at my land. A nice orchard of more than 25 trees is nowhere to be seen. Thousands of dollars worth of timber is under the rocks and trash pushed down by machinery. On tops of the mountains all you can see are acid wastes and stagnant pools. . . . With the heavy rains beating down these bare hillsides what will hold the soil? . . . The poor people in the valleys will be covered with the filth leaving no farm land to grow crops.[17]

Strip mining made arable land useless, dumped acid and silt into waterways, and contaminated local water supplies. Even though it also provided jobs, grassroots campaigns to regulate or ban strip mining grew up in many areas in the years after World War II.

The Movement against Strip Mining

During the first part of the 20th century, environmentalism was often seen as a movement by the elite, aimed at preserving spiritual, aesthetic, and recreational values at the expense of the economy. This view was never wholly accurate. George Perkins Marsh in the 19th century and Gifford Pinchot in the earlier 20th emphasized the need to conserve the nation's forests in order to be able to use their resources in the future.[18] However, John Muir's campaigns to save the redwood trees and preserve the natural wonders of the Yosemite and Hetch Hetchy Valleys as temples of nature, or Bob Marshall's insistence that wilderness was essential to the soul of the nation, even though only the most virile would be able to make use of them, certainly reinforced this impression.[19]

The movement to stop strip mining, in contrast, was a grassroots movement of middle and lower class people, primarily in rural areas. While these people, too, cited aesthetic and spiritual values, including the Christian duty of stewardship, their primary emphasis was on the threat strip mining posed to their livelihoods and to their right to their own property. Montrie concludes that "the most important cultural lens refracting abolitionist arguments, and significantly influential in shaping grassroots demands for regulatory laws, was an American tradition of veneration for small private property.[20] For those involved, the jobs vs. the environment dilemma simply did not apply. Environmentalists have often sought with mixed success to build coalitions with local logging and fishing communities around similar arguments that excessive exploitation will destroy local livelihoods;[21] here, it was not a matter of environmentalist leaders building a coalition, but of the spontaneous development of environmental concerns from the grassroots. Montrie details the development of grassroots movements against strip mining in Ohio, Pennsylvania, Kentucky, and West Virginia.

As the movement grew, and state governments proved resistant to controlling strip mining, the demand arose for the federal government to get involved. This in turn led to more involvement by national environmental groups. From the local point of view, such involvement was a two-edged sword. The national groups brought knowledge of and connections to Washington politics as a resource, but they also brought a habit of compromise and incrementalism. Their involvement helped bring about the eventual passage of the Surface Mining Control and Reclamation Act (SMCRA) in 1977, but in a form that had been so badly weakened in Congress that many local environmentalist groups urged President Carter to veto the bill. However, Carter signed it into law in August.[22] Despite the weaknesses of the law, strip mining was now subject to federal regulation.

The Legacy of the Anti–Strip Mining Movement

The SMCRA had many weaknesses. It set federal standards but left implementation to the states, whose officials tended to side with the mine operators; the standards themselves were thought by many to be too weak; and the new law did little to stop the growing new practice of mountaintop removal. Montrie provides case studies of two grassroots organizations, Save Our Cumberland Mountains (SOCM) in Tennessee and Kentuckians for the Commonwealth (KFTC) in Kentucky, that worked to secure stronger enforcement of the SMCRA. SOCM had begun its life campaigning to ban strip mining altogether, citing the inevitable damage to the land, the loss of property rights by farmers, and the loss of jobs in comparison with deep mining. Following passage of the SMCRA, they turned their attention to the enforcement of the law. As they did so, SOCM activists were subject to violent attacks by coal operators, including the burning of the houses of Sam and Roberta Baker, John Johnson, and Millard and Mable Ridenour (actions for which no one was ever charged).[23] The tradition of mine operator lawlessness dating back to the 19th century proved difficult to overcome. However, they did block some proposed new mines and eventually won a struggle to have enforcement in Tennessee federalized because of the perceived collusion of the state government with the mining companies.

Meanwhile in Kentucky, KFTC focused its efforts on two goals: making mine owners pay a fairer share of taxes and amending the state constitution to limit the impact of the broad form deed. They eventually made

progress in both areas. The state supreme court ruled that the prior practice of taxing coal deposits at a lower rate than other property violated the
state's constitution, forcing operators to pay much higher taxes from 1990
on;[24] and in 1993, after some initial defeats, the group pushed through an
amendment to the state constitution banning the issue of a permit for
strip mining without the express consent of the owner of the surface
rights to the land, regardless of any mineral rights deed.[25] These victories
restricted the scope of strip mining in the future, but they did nothing to
restore the land and waterways that had been devastated by past strip
mines throughout Eastern Kentucky.

Recently the coal operators have turned to a new method: mountaintop
removal. This is just what it sounds like; the upper parts of mountains are
removed with explosives and heavy equipment to get access to the coal
beneath. Current technology makes it feasible to extract coal from as
much as four hundred feet below the surface. Since few farms are actually
on mountaintops, the operators do not have to deal with farmers; however, much of the overburden is disposed of as "valley fill;" that is, it is
pushed over the edge of the mining site to tumble down the mountain
side, with accompanying damage to the timber and waterways below. The
practice has grown rapidly; West Virginia issued mountaintop removal
permits for only three hundred acres in 1973, but over twelve thousand
acres in 1997. Perhaps because the technique of mountaintop removal
was rare at the time the SMCRA was passed, it is not dealt with explicitly
in the law. Operators are required to restore strip mining sites to the
"approximate original contour" (AOC) of the land, which might be interpreted as meaning that the mountaintop would have to be rebuilt. However, regulators have tended to go along with the operators' claim that
such restoration is not required. One study found that restoration of former mountaintops to rolling fields described as "fish and wildlife habitat"
or "timberland" was frequently approved. The grassroots movement
against mountaintop removal remains at an early stage.[26]

Lessons of the Strip Mining Struggles

Past conflicts over strip mining exhibit a few common themes that may
be relevant to the coming conflict over greenhouse gas emissions. These
include the lawlessness of the operators, the role of rural communities
and their grassroots organizations, and the ambiguous position of organized labor.

From the beginning of industrial coal mining in America, the mine
operators have been accustomed to having their way with the law.

Through a combination of political influence and brute force, early operators were able to acquire land titles against the will of the occupants, to suppress labor activism with legal and extralegal violence, and to destroy farms and homes in order to get at the coal beneath. This history may be part of what gives rise to the willingness of present-day coal owners to go all out in order to prevent any regulation of greenhouse gases.

The resistance to strip mining has been rooted in coalitions of farmers concerned about their land, sportsmen concerned about the health of fish and game populations, and rural communities concerned about the loss of the tax base to finance local education and other public services. As their campaigns against strip mining grew, they drew in national environmental organizations as allies. However, the resulting alliances have sometimes been problematic, and local groups have sometimes felt betrayed by policy compromises negotiated nationally. Strip mining illustrates both the possibilities of building cross-class environmental coalitions and the problems that may possibly arise in doing so.

One of the largest such problems is the role of labor. On the one hand, labor unions have often clashed violently with management, as discussed above with regard to Pennsylvania, but also in the West, as in the case of the Ludlow Massacre.[27] On the other hand, unions can sometimes side with employers in the belief that doing so will preserve jobs in the industry. With regard to strip mining, the UMW has come down on each side at different times, sometimes using support for strip mine regulation as a bargaining chip with the employers, and sometimes opposing it for the sake of jobs. All three of these themes will be relevant to the conflict over coal and climate change.

Conclusion: The Politics of Coal and Climate Change

The climate movement now seeks to eliminate the burning of coal and, as a first step, to close all coal-fired electric plants in the United States.[28] In many ways this is a completely new variety of conflict. First of all, it involves the continued survival of human civilization. This point is not trivial; one might expect that, as understanding of climate change grows both among the general public and in the corporate community, there would be an accompanying sense that something has to be done, regardless of the immediate interests of particular business sectors. However, we are not yet at the point where such general human interest dominates the policy process.

Even short of saving the world, the lineup of political forces is somewhat different on climate issues. Other sectors of the movement are seeking to build a blue-green alliance around green jobs and the potential to develop an export industry in photovoltaics and wind-turbine components, as well as in adaptation to (as opposed to mitigation of) climate change. As local governments respond to the desires of their citizens to play a role on climate, the relation of the progressive federal government to conservative states has been partially reversed.

That said, there remains much to learn about the politics of climate from the politics of coal in these earlier contexts. Unless the interests of workers can be aligned with the interests of saving the climate, and the feelings for the land of people in local communities can be linked with the science-driven approach of classical environmentalism,[29] it will be difficult to find a climate solution soon enough for it to be useful.

CHAPTER THREE

Coal and Air Pollution

The damage done by coal smoke inspired the first attempts to regulate air pollution. However, coal was such a valuable energy source that it was never thought possible to limit its use. Instead, efforts were focused on technological improvements to remove pollutants from coal's exhaust stream. Only in the late 20th century did it begin to seem either possible or desirable to replace coal with cleaner energy sources. By that time, the earlier attempts to regulate air pollution had created a policy structure that was ill suited to the task.

People have always known that the smoke from burning wood or coal was unpleasant to breathe. Long before the chemistry of combustion was known, even the simplest of cold-climate buildings was designed with a way for the smoke from fires to be vented to the outside; the tipis of the Dakota had smoke holes, as did the igloos of the Inuit. As coal powered the early days of the Industrial Revolution, it also covered areas near factories with soot, and there was a general understanding that smoke, especially smoke from coal, caused respiratory diseases such as tuberculosis and asthma.[1] However, early efforts at "smoke abatement" were based on the assumption that the smoke itself was the problem, with the seriousness of the toxicity calibrated with the darkness of the smoke. "Smokeless grates" were designed to recirculate coal smoke for further combustion, and the slogan "burn your own smoke" was propagated.[2] Only in the mid-20th century did it become generally known that other, invisible pollutants, especially the oxides of sulfur and nitrogen, were even more toxic than the components of visible smoke.[3] The need to control these chemical pollutants was brought home by two mid–20th-century air-pollution disasters, one in London, and one just outside of Pittsburgh in Donora,

Pennsylvania, as well as by a massive air-pollution problem in Los Angeles.

Air Pollution Disasters

The London and Donora disasters had similar causes; in each case a thermal inversion, in which a layer of warm air trapped a colder layer beneath it, caused the exhaust of household and industrial fires to reach unusually high concentrations. The health consequences were severe, with about four thousand deaths in London and twenty in Donora, during the duration of these pollution emergencies. In each case, many more died later. Although these later deaths were not as sudden and dramatic, epidemiologists demonstrated that the death rate was significantly higher following the inversions than at other times, and medical doctors were able to trace the etiology of individual deaths back to the polluted air.

Donora

Donora, Pennsylvania, is one of the industrial suburbs that grew up around Pittsburgh at the time when the latter was the heart of the American steel industry. Donora is situated at the base of six-hundred-foot hills along the Monongahela River. In the 1940s, it was home to steel mills, coke plants, and a very large zinc smelter, all of which were powered by burning coal.[4] Smoke was a constant presence, so much so that Devra Davis, who grew up in Donora, gave her book on pollution the title *When Smoke Ran Like Water*. Smoke and soot were so common that local residents did not think of them as pollution, just as the way things were.

On Tuesday, October 26, 1948, that smoke became deadly. A thermal inversion trapped the exhaust of the mills within the valley; as the mills kept running, the smoke and fog got thicker. Emissions from the zinc plant included highly toxic fluoride gas; Davis later showed that most of those who died lived within the plume of emissions from this plant and most likely were killed by the fluoride.[5] The zinc plant was not shut down until 6:00 a.m. Sunday, shortly before rains came and the inversion lifted. By that time, twenty people had died; another fifty were to die over the next few months. Approximately six thousand Donorans suffered adverse health effects.[6]

The Donora disaster drew national attention. Newscaster Walter Winchell announced on his national broadcast that "people dropped dead from a thick killer fog that sickened much of the town."[7] Most government authorities regarded the cause as just that: a "killer fog" or "bad air."

Nevertheless, a coalition of prominent local business figures, led by Richard King Mellon, worked with Mayor David Lawrence to bring about a "Pittsburgh Renaissance," including ridding the city of the smoke and soot that defined its reputation. Mellon was one of the richest people in America and controlled several large companies, a status that gave him considerable ability to get things done; for example, when the Pennsylvania Railroad tried to block the state from granting air pollution control powers to Allegheny County, Mellon informed them that "the companies he controlled would probably switch their freight haulage to other, competitive lines," whereupon the railroad withdrew its opposition.[8] Eventually the Pittsburgh air pollution office was merged into Allegheny County's, and the latter established a largely voluntary program of working with industry to reduce air pollution. Smoke reduction was helped considerably by the collapse of the American steel industry, but other accomplishments were limited.

Meanwhile, the desire to explain what had happened in Donora inspired a number of scientific studies of the toxins present in the exhaust from the local mills and the effect of those toxins on human health. Although it took years for the results of such studies to win wide acceptance, they opened the door to the air pollution regulatory regime eventually launched under the Clean Air Act.

London

Coal was brought to London beginning in the 12th century as ballast in otherwise empty cargo ships returning from the North. It was first used primarily in industry, particularly for the production of lime. A commission to study the resulting air pollution and to find a solution for it was created in 1286, and continued to operate for twenty years, although its recommended solutions failed to have much effect.[9] Use of coal for residential heating was inhibited by the lack of adequate chimneys; wood could be burned with less toxic smoke, and charcoal was essentially smokeless. However, as the population of the city grew and wood became scarce in the vicinity of London, houses began to be fitted with chimneys and to switch to coal as a source of heat. The drastic loss of population to the Black Death (1348–1351) allowed a temporary return to firewood, but continued population growth soon required a more plentiful source of fuel once more. Coal imports to London increased over thirtyfold, to more than 360,000 tons a year, between 1580 and 1680.[10] Residential use of coal was at first a mark of poverty, disdained by the aristocracy, but all that changed with the accession of James I, who had been accustomed to household coal burning while he was in Scotland and continued the practice in London.[11] The king's example made heating with coal fashionable. By the 19th century, London had become the world leader in burning coal, and as a result developed its famous "pea soup" fogs, which were featured in the novels of Charles Dickens and Arthur Conan Doyle and painted by a visiting Claude Monet.[12]

The smoke and fog mixture of London was always considered a nuisance, and there were occasional incidents that suggested that this mixture was a danger to health. In 1873, prize cattle brought to the city for Cattle Show Week suddenly developed bronchitis and had to be slaughtered, and there was some speculation that they may have reacted to the smoke.[13] As early as the 17th century, John Evelyn, a founder of the Royal Society, had proposed that coal-burning commercial establishments be moved to the periphery of London.[14] However, Evelyn's recommendation was not taken up, and no effective regulation of coal smoke was adopted for centuries. The killer fog of 1952 changed this.

On December 5, 1952, a blanket of cold air descended on London, both motivating Londoners to stoke up their coal fires and creating the inversion that prevented the smoke of those fires from escaping. The inversion continued through December 9, as the level of various pollutants rose.[15] Within twelve hours of the increase in pollution, the number of deaths began to rise as well. The deaths were not as dramatically sudden as those in Donora, but there were many more of them. The mortality level remained elevated until December 20; public health researchers calculated that four thousand people had died from causes attributable to the fog by that date.[16] Two weeks later the number of deaths rose again, less acutely but for a longer period; there were an additional eight thousand extra deaths by the end of February 1953.[17] Although government ministers at first saw no need for legislation, maintaining that local governments already had the power to control smoke, popular pressure led to the passage of the Clean Air Act of 1956. The act was weak in several ways; it applied only to smoke and specified that industry be required only to use the best "practicable" means of smoke control. It did establish a basis for further regulatory action, however. Meanwhile the United States was observing a less acute but longer lasting air pollution crisis in Los Angeles.

Los Angeles Smog

During the late 19th century, Southern California became known for its pleasant climate and pristine atmosphere and was the site of a number of sanitaria where people with tuberculosis could go for recovery, as well as resorts in such places as Pasadena.[18] However, the cleanness of the air was due solely to the lack of pollution sources; the local geography and climate were such as to make temperature inversions, which prevented any pollution from dispersing, extremely common. The Pacific Ocean cooled air near the surface, which was then held in the Los Angeles Basin by the surrounding mountains and kept from rising upward whenever the air at higher elevations was warmer, as it often was.[19] As early as 1542, the explorer Juan Rodriguez Cabrillo had observed the results of such an inversion, which made the smoke from campfires hang in the air over San Pedro Bay, which he called the "Bay of Smokes" as a result.[20]

The importance of tourism, and the city's pride in its climate, led to the passage of a smoke control ordinance in 1905—and others in 1907, 1908, 1911, 1912, and again in 1930. These laws, and the smoke inspectors they authorized, achieved some improvement; the replacement of coal by oil and natural gas, once the latter were found in California, probably helped even more.[21] But the 1940s brought an ominous new development, the frequent occurrence of dark low-lying clouds of what became known as "smog." This term, a portmanteau of "smoke" and "fog," had originated in London, where it described the local pollution aptly, and it was brought into scientific usage in 1905 by Henry des Voeux of the Coal Smoke Abatement Society.[22] In Los Angeles, it was completely inaccurate, as neither smoke nor fog was present in significant amounts. Nevertheless the term stuck.

Unlike London or Donora, Los Angeles did not experience deadly waves of sudden death caused by air pollution, but things were bad enough. Smog stung the eyes, reduced visibility, and led to increased respiratory disease (although the last phenomenon was not understood until years later). Citizen outrage led to a number of efforts by both city and county government to get smog under control, with little initial success. In 1946, the leading local newspaper, the *Los Angeles Times*, launched a campaign to do something about the smog. The decision to do so was made by editor Norman Chandler, whose family owned the newspaper, at the urging of his wife, Dorothy Buffum Chandler, who later reported that she had been driving back to the city over the mountains, had seen the layer of smog over the city, and "I just went into the office and said to Norman, 'Something has to be done.'"[23]

Chandler assigned a full-time smog editor and hired an outside expert to study the situation and make recommendations. Raymond R. Tucker had been the smoke-control officer for the city of St. Louis, where he had achieved notable success. The problems of St. Louis had mostly been due to coal combustion in industry and households, and most of Tucker's twenty-three recommendations were based on smoke control in St. Louis and other eastern cities; but he also recommended seeking state legislation to authorize the creation of a countywide air pollution–control district.[24]

Los Angeles County contains a large number of independent municipalities, some of them containing little population but much industry. The City of Vernon, for example, incorporated in 1905, was the workplace for 10 percent of the region's factory workers, but had a population in the 1940s of 417, many of whom were employees of the City of Vernon. The city's motto, displayed on its seal, is "exclusively industrial."[25] Such small industrial cities had no interest in controlling air pollution, but their status made them exempt from county regulation. While many of these small municipalities resisted county control of emissions, eventually all agreed to give up this amount of home rule; after a protracted lobbying struggle, the bill was signed into law by Governor Earl Warren in June 1947, and the Los Angeles Air Pollution Control Board (APCB) was born.[26]

Resisting industry pleas that action be delayed until science had determined the true source of smog, the APCB under its first director, Louis McCabe, set out to control all sources of pollution to the limits of technological feasibility, while simultaneously launching a large research program to reduce those limits. The agency succeeded in banning household incinerators and reducing industrial smoke emission to a great extent. The smog continued, albeit with a slight reduction in frequency. It is now generally understood that the principal source of smog was not stationary sources or backyard incinerators, but rather the automobile, but it took a long time and considerable effort for this to be believed. Raymond Tucker's report on smog had included the heading "Automobiles Absolved from Most Blame," and McCabe had asserted in 1949, "Neither should folklore be encouraged that will place the onus of metropolitan area atmospheric pollution on the automobile, without proof."[27]

Proof was eventually provided by Arie Haagen-Smit, a chemist at the California Institute of Technology, who had previously specialized in the study of food flavors. Haagen-Smit was able to use his lab equipment to analyze the composition of smoggy air, and he eventually realized that the damaging components of smog were not emitted directly but rather formed through chemical reactions in the atmosphere energized by sunlight: photochemical reactions. His theory was resisted by industry, and by many scientists, and his early articles about it were rejected by scientific journals, but his argument came to be accepted after he presented it with comprehensive supporting data to the Second National Air Pollution Conference in 1952.[28]

Air pollution in Los Angeles proved to have little relation to coal. Nevertheless, it is an important part of the story of how the burning of coal came to be regulated. The pollution in and around Pittsburgh was local in origin, and ultimately could and would be controlled locally—at least so it seemed in the early 1960s. Los Angeles County and the state of California also attempted to control automobile emissions locally, and they did achieve some success. Ultimately, however, they did not have the power to compel the manufacturers to develop new technology. At the extreme, the manufacturers could have closed their plants in California and simply imported automobiles from elsewhere. Short of banning cars from the state or city outright, the only practical solution was national regulation of emissions. Thus it was the battle against smog in Los Angeles that led most directly to federal air pollution regulation, regulation that came to encompass coal as well.

The Battle for Clean Air

By the late 1940s people concerned with air pollution were beginning to realize that the problem was beyond the capacity of local and state governments to handle. However, the proper federal role was thought by almost everyone to be limited to providing technical expertise, and perhaps financial support, to the states. In 1949, at the request of several federal agencies that had been meeting informally to discuss air pollution, President Truman authorized an official interdepartmental committee and a technical conference on the matter. His directive included a limitation:

> I do not contemplate that the deliberations of the Interdepartmental Committee and the Conference will result in the creation of programs which will commit the Federal Government to material expenditures from an already heavily burdened treasury, since the responsibilities for corrective action and the benefits are primarily local in character.[29]

This perspective was widely shared at the time, not only for the fiscal reasons cited by Truman but out of a conviction that American federalism put the problem of air pollution under state jurisdiction. The first significant federal legislation on the subject, the Air Pollution Control Act of 1955, authorized federal research and technical assistance to the states, but added:

> It is hereby declared to be the policy of Congress to preserve and protect the primary responsibility and rights of the States and local governments in controlling air pollution.[30]

The primacy of state and local government is deeply rooted in American legal and political philosophy. The philosophers of the Enlightenment had argued that individual rights to life, liberty, and property were an essential limit on the power of rulers; if these rights were not preserved, tyranny was likely to result.[31] This view of property rights was embedded in the Constitution, both in the Bill of Rights and in the limited grant of powers to Congress in Article I; Congress was meant to have no powers other than those listed, an understanding later made explicit in the Tenth Amendment. Government was also understood to have the basic power and responsibility to protect the public health and safety through "police powers," which could sometimes override the protection of rights. Individuals could be quarantined to prevent the spread of disease, or buildings knocked down to prevent the spread of a fire, for example. However, the Constitution was intended to restrict such powers to state and local government exclusively.

Of course state governments could also violate rights and become tyrannical, and the U.S. Constitution did place some limits on their powers: they could not issue currency or annul contracts, for example. Most state constitutions also protected basic rights, and the Fourteenth Amendment eventually gave a federal guarantee that states could not violate them. But the understanding of property rights also evolved over time to meet changing conditions, and such practices as zoning, building codes, and mandatory immunization became accepted.[32] When California set out to require pollution controls for automobile engines, it faced many practical difficulties, but there was no doubt about its legal authority to do so. However, federal authority in this area was less clearly established.

For the most part federal regulatory power is based on the commerce clause in Article I of the Constitution, which gives Congress the power "to regulate Commerce with foreign Nations, and among the several States, and with the Indian Tribes," among other things. While this clause was initially limited in its application to regulation of the terms of trade, it has gradually evolved to meet the needs of a changing society, so that it now extends to limitations on child labor, a federal minimum wage, and health regulations, for example.[33] The question facing Congress, then, was whether air pollution was part of interstate commerce. Prior to 1960, the consensus had been that it was not. But some were beginning to disagree, for two reasons. Only the first of these was directly relevant to coal: since air pollution did not stop at state boundaries, states could suffer from pollution originating in other states, which only the federal government had the power to control. The second reason pertained to automobiles; however, by helping establish the logic of federal regulation, it indirectly affected coal as well.

As it became clearer that motor vehicles were one of the major sources of air pollution, it also became clear that no single state was in a position to require a redesign of the automobile to make it less polluting. Rep. Paul F. Schenck (R-OH) began in 1958 to introduce bills to mandate that the surgeon general set standards for the amount of unburned hydrocarbons that an automobile could emit without endangering public health and that vehicles not meeting this standard be banned from interstate commerce. The bill was given a hearing at the subcommittee level, where it was supported by a representative of Los Angeles County but opposed by both the auto industry and by Marian B. Folsom, Secretary of Health, Education, and Welfare (HEW), to whom the surgeon general reported, on the ground that the knowledge required to make such a determination simply did not exist; the bill went no further. The next year Schenck introduced a new bill, which mandated research and specified that regulations would be issued only when enough research had been done to justify them, but it too failed, with Arthur Flemming, who was now Secretary of HEW, asserting in a letter to the House Committee on Interstate and Foreign Commerce that "the preponderance of air pollution problems in the United States is intrastate in character." Schenck finally proposed a bill that simply mandated research into the health effects of air pollution, with a report to be made to Congress within two years. This bill was passed and signed into law by President Eisenhower on June 8, 1960. The resulting report was to become the first step toward federal regulation of air pollution.[34]

Clean Air Act of 1963

In response to Representative Schenck's bill, the surgeon general sent a report to Congress in June of 1962 titled "Motor Vehicles, Air Pollution, and Health." The report stated flatly, "The presence in the air of pollutants originating from motor vehicles has been demonstrated beyond doubt," and endorsed Haagen-Smit's conclusion that some of these pollutants were intensified by photochemical reactions. However, the report also concluded that the evidence of health and economic harm had been demonstrated qualitatively but not quantitatively, so that it was not possible to recommend specific air quality goals or emissions limits. In response, the

Senate Committee on Public Works added a provision to the Clean Air Act of 1963 to direct the surgeon general to do further research with a view of obtaining such quantitative evidence.[35]

Most of the law dealt with stationary sources of air pollution, including coal and oil-burning factories and power plants, along with various direct emissions from chemical plants. It continued the program of federal grants for the development of local air pollution–control programs, and for research, that had been established in 1955. However, it also authorized a federal role in pollution abatement in cases where pollution in one state affected the health or welfare of people in other states.

A federal role in enforcement was opposed by much of industry. It was also controversial within the federal bureaucracy and historically had had little support within Congress; only Representative Schenck had proposed federal regulation of air pollution, and that only for motor vehicles. Most importantly the chair of the House subcommittee responsible for air pollution legislation, Rep. Kenneth Roberts (D-AL), had repeatedly affirmed his belief that air pollution should be handled by state and local governments, with only research assistance and financial support coming from the federal level. However, the idea of federal regulation was supported by the organized "urban lobby": the United States Conference of Mayors (USCM), the American Municipal Association (AMA), and the National Association of Counties. Several other local governments had come to the same conclusion as Los Angeles, that neither state nor local governments could control air pollution without federal help. Hugh Mields Jr., who served first as assistant director for federal affairs at AMA and then as associate director of USCM beginning in 1962, was to play a key role in developing legislation.[36]

The newly created (as of 1960) Division of Air Pollution, headed by Vernon MacKenzie, was placed in the Bureau of State Services as part of the Public Health Service (PHS), which itself was part of the Department of Health, Education, and Welfare (HEW). While MacKenzie favored a federal enforcement role, the PHS as a whole was opposed to it, partly because they considered research to be their fundamental mission and feared that having any enforcement power would jeopardize their relationships with state health agencies, and partly because they thought giving them enforcement powers would jeopardize passage of the bill, which was mainly an authorization for more research funds. Federal enforcement was also opposed by the Bureau of the Budget as well as some other agencies. However, MacKenzie went outside the PHS to work directly with Dean Coston, special assistant to the Assistant Secretary for Legislation of HEW, Wilbur Cohen. Cohen was scheduled to fly to Palm Beach,

Florida, in December 1962 to discuss the department's legislative priorities with President Kennedy, who was vacationing there. Coston won Cohen's support for a presidential endorsement of federal enforcement powers, and Kennedy agreed to do so. After more bureaucratic back-and-forth, in which Mields also played a role, Kennedy's message on health on February 7, 1963, included a recommendation that the PHS be authorized, among other things, "to take action to abate interstate air pollution, along the general lines of the existing water pollution control enforcement measures."[37]

Meanwhile, Representative Roberts had changed his mind about federal enforcement. A second killer smog in London in 1962 had left 340 dead, and Roberts had also become increasingly concerned with the air pollution problems in Birmingham, which adjoined his district. After a series of meetings with Coston in February 1963, he agreed to introduce a bill providing for federal action on interstate air pollution abatement, as well as increased funds for grants to state and local governments.[38] Roberts's subcommittee held two days of hearings in which the only outright opposition to federal enforcement came from the National Association of Manufacturers (NAM). The NAM spokesperson quoted an earlier Roberts speech in which he had said that pollution should be dealt with locally; Roberts replied, "The wise man changes his mind and the fool never does." Other industrial representatives focused on specific changes they sought in the bill. No one representing the coal industry testified in the House hearings. Roberts made a few adjustments in the bill, which went on to House passage on July 16, 1963, by a vote of 272–102.[39]

Senate action on the Roberts bill was affected by two events. First, the two senior Democrats on the Public Works Committee, which had jurisdiction over air pollution, had died that winter; one of these, Robert Kerr (D-OK) had been a powerful voice against any federal role in regulating pollution. Pat McNamara (D-MI) became the new committee chair, and appointed Edmund S. Muskie (D-ME) to chair a special subcommittee on air and water pollution. Muskie was to become the Senate's leading expert on pollution.

Second, Kennedy's first Secretary of HEW, Abraham Ribicoff, had resigned in order to run successfully for the Senate in his home state of Connecticut. As a new senator, Ribicoff introduced a strong air pollution bill, providing for a federal role in enforcement for intrastate as well as interstate pollution. He did not receive the seat he had requested on the Public Works Committee, so he had only limited involvement in consideration of the bill, but by calling for a federal role in even intrastate pollution, his bill influenced the legislative outcome, and it was his bill that was used as the framework for Senate amendments.[40]

Industry associations, including the National Coal Association, were more involved in the Senate process than they had been in the House, but they had only a small effect. Both Muskie and the ranking minority member of the Public Works Committee, Sen. Caleb Boggs (R-DE), wanted a strong bill, and they had the votes to get one. The only significant concession to industry was to grant a request from the Manufacturing Chemists Association that the authority the bill gave to the Secretary of HEW to order a company to submit a report on a pollution problem be limited to reports based on data already in their possession, rather than forcing them to do expensive new research. The bill passed the Senate on November 19, both houses approved the conference report on December 10, and President Johnson signed the Clean Air Act of 1963 on December 17.[41]

In theory, the federal government now had significant new power to control air pollution. However, the procedure for federal action was long and cumbersome; the Secretary of HEW, either on his own initiative or at the request of a state, could call a conference on a particular case of interstate or intrastate air pollution, the conference would make recommendations, the secretary could then choose to issue an abatement order, and if needed could ask the attorney general to seek legal action. Moreover, the responsible office, now called the National Center for Air Pollution Control, was small, underfunded, and still buried five levels down in the structure of HEW; by 1966, only eight interstate abatement actions had been initiated.[42] President Johnson called for the federal government to "vastly expand the fight for clean air," and called for federal emissions controls for some industries. This was to become one of the major issues as Congress debated strengthening of the Clean Air Act over the next few years.[43]

Development of the Clean Air Act after 1963

With the enactment of the Clean Air Act of 1963, the federal government had the power to regulate air pollution both from motor vehicles and from stationary sources; coal, in the form of coal-burning electric generating plants, fell into the latter category. The specifics of such regulation were left to be worked out by the National Center for Air Pollution Control, part of the Public Health Service. To implement the new law, the Center had to perform a number of difficult tasks: determine which pollutants were harmful, calculate what atmospheric levels of these pollutants could be tolerated as not dangerous to health, decide how much emissions from polluting industries had to be reduced in order to attain desirable atmospheric levels, and finally enforce those standards by measuring the emissions from individual sources and ordering reductions where required. The Center had neither the budget, the labor power, the technical expertise, nor sufficient bureaucratic leverage to accomplish these tasks. Meanwhile, environmentalist consciousness was on the rise and the demand for action on air pollution was increasing. The surprising success of the first Earth Day in 1970 added to the pressure for federal action. As a result, the Clean Air Act was amended in 1965, 1967, 1970, and 1977, with each change increasing the role of the federal government.

The basic mechanism for air pollution in control was in place with the passage of the 1970 Clean Air Act, with a few refinements added in 1977. The amendments called for the establishment of National Ambient Air Quality Standards (NAAQS) by the federal government. These standards were to be based on the protection of human health; the responsible agency was to determine which pollutants presented a potential threat to health, and at what level of pollution this threat was significant. Evaluation of six substances, designated as "criteria pollutants," was mandated in the language of the act. These were carbon monoxide, ground-level ozone,[44] nitrogen oxides, particulate matter, sulfur dioxide, and lead.[45] The agency was also mandated to evaluate the health effects of additional pollutants and to set air quality standards for those pollutants if merited. The country was divided into 246 air quality regions; if a region failed to meet the level set by the NAAQS, the corresponding state had to develop a plan to bring the region into compliance and submit that plan for federal approval.

In addition to the ambient air quality standards, the law also provided for emission standards. These were most important for mobile sources of pollution (i.e., motor vehicles). Motor vehicle emissions standards were the subject of dramatic conflict as the auto manufacturers repeatedly sought to have their implementation delayed, arguing that strict enforcement would make new cars unaffordable and devastate the auto industry and the overall American economy. Since these conflicts did not involve coal, they need not concern us here.[46] However, there were also emissions standards for stationary sources, which were primarily coal- or oil-burning electric plants, along with some chemical manufacturers. Since it was more difficult to retrofit pollution controls in existing plants, which in any case had a limited lifespan, the emphasis was on emissions standards for new plants, known as New Source Performance Standards (NSPS), which were to be set by the federal government. However, in cases where an air quality region failed to meet the NAAQS, existing plants might also be regulated as part of a State Improvement Plan.

While the bill was working its way through Congress, a significant administrative change occurred. President Nixon issued a reorganization plan to create a new agency, the Environmental Protection Agency (EPA), and to transfer most environmental programs to it; the National Center for Air Pollution Control was included in the reorganization. As a result, air pollution was now at the second level of an important independent agency, rather than the fifth level of a department. However, the new agency continued to be understaffed in proportion to its mandated responsibilities, so much so that progress in setting standards continued to be slow. Impatience with this slow pace continued to induce environmentalist members of Congress to seek to enact specific standards by legislation, rather than leaving them to be determined by EPA's evaluation of the scientific evidence. A controversy over such legislative specification of standards during consideration of the Clean Air Act amendments of 1977 led to the first major involvement of the coal industry in trying to influence air pollution policy, specifically with regard to the control of sulfur dioxide emissions.

The coal industry had not been much involved in the early struggles over air pollution control. Although the combustion of coal was one of the major sources of pollution, most people in policy-making circles assumed that the answer was to remove the pollutants from the exhaust, not to burn less coal. However, this lack of involvement began to change as pollution from sulfur became a larger concern. Coal deposits contain varying amounts of sulfur, which is expensive to remove. When the coal is burned, the sulfur burns as well, producing sulfur dioxide; this in turn can combine with water (in the air or in lakes) to form sulfuric acid. Awareness of the corrosive effect from acids in coal smoke goes back to the time of John Evelyn; however, neither the extent nor the harmfulness of acid deposition had been measured. This began to change in the 1970s with the publication of a series of studies from the Hubbard Brook Experimental Forest in New Hampshire showing that the waters in this rustic area, far from the sources of pollution, had been growing more acidic for at least twenty years.[47]

An early remedy to pollution from coal plants had been to build taller smokestacks so that the contaminants would be dispersed over a wider area, an approach described colloquially with the slogan "the solution to pollution is dilution." However, given the direction of the prevailing winds, this amounted to making rural northern areas in other states and countries suffer the damage caused by activity in industrial areas far away and thus became unacceptable. It became clear that sulfur would have to be removed from the exhaust stream, not just diluted.

Sulfur emissions can be reduced by three different methods. Coal can be washed before burning, which removes some of the sulfur content; devices known as "scrubbers" can be used to remove sulfur from the exhaust fumes after coal has been burned; or coal with high sulfur content can be replaced as a fuel by coal with less sulfur. The last solution is obviously the easiest, except that low-sulfur coal is mostly available in the mines of Wyoming and Montana, further from Eastern power plants than the high-sulfur coal of Appalachia. A move by Eastern power plant operators to switch to coal from the West presented a serious economic threat to both the owners of the Eastern coalfields and the miners who worked in them. The Eastern coal interests responded to this threat by proposing that government mandate a specific technology rather than a quantitative limit on sulfur emissions. Stack scrubbers were capable of reducing sulfur emissions by about 70 percent—at least when they were working well. Of course, how much sulfur remained in the exhaust after scrubbing would depend on how much sulfur was present in the coal before combustion. However, if all power plants were required to use scrubbers, and if using scrubbers was considered a sufficient remedy for pollution, Eastern coal would not be at a disadvantage in the market.

Meanwhile, the major environmental organizations were concerned with another issue, nondeterioration of regions that still had very clean air, most of which were in the rural West. The air in such areas could become more polluted without violating the NAAQS, and this was thought undesirable. Driven by this concern, the environmental lobby supported the scrubber requirement so that the emissions from Western power plants, which were going to burn low-sulfur coal anyway, would be even cleaner. The alliance of Eastern coal interests with major environmental organizations was successful; as a result, Eastern air was dirtier than it might have been while Western air was cleaner, a state of affairs that continued until passage of the next major amendments to the Clean Air Act in 1990.[48]

Prior to the 1980 election, environmental policy had been largely bipartisan. In the House of Representatives, the major conflict on air pollution matters was between Democrats John Dingell of Michigan, who sought to protect the automobile industry from having to install expensive antipollution devices in cars, and Henry Waxman of California, who wanted stricter emissions requirements on cars in order to help Los Angeles solve its problem with smog.[49] In the Senate, Robert Byrd of West Virginia, Democratic floor leader from 1977 to 1989, defended the

interests of high-sulfur Eastern coal. Although there were partisan disagreements about methods and costs, many Republicans supported
environmental protection. Most importantly, Republican presidents
Nixon and Ford wanted to be seen as pro-environment; Nixon's appoin-
tee as first head of the EPA, William Ruckelshaus, was generally seen as
an effective defender of the environment.[50]

This bipartisanship began to disappear with the election of Ronald
Reagan to the presidency in 1980. Reagan appointed James Watt as Secretary of the Interior and Anne M. Gorsuch (later Anne Gorsuch Bur-
ford) as administrator of the EPA; both of these figures were openly
skeptical of environmental regulation, which they believed to be unfa-
vorable to economic development, and they evoked the distrust of many
Democratic members of Congress. Although each was later replaced by
a more moderate figure—Reagan brought back Ruckelshaus in an
attempt to restore the credibility of the EPA—their actions while in
office helped increase partisan polarization over the environment. Environmentalists were unhappy with the lack of progress toward attaining
cleaner air, but the partisanship within Congress and the opposition of
the president prevented the passage of any legislation to strengthen the
Clean Air Act.

The door to such a law was reopened in 1989, with two crucial changes
of personnel. George H. W. Bush, who had proclaimed that he wanted to
be "the environmental president," succeeded Reagan, and George Mitchell
replaced Robert Byrd as Majority Leader of the Senate. Although they
disagreed on many details, both Bush and Mitchell wanted a stronger
clean air law and were willing to accept some compromises in able to get
such a law. At the same time, Congressional Democrats continued to distrust the EPA. As a result, the Clean Air Act of 1990 went unusually far in
specifying what should be done, rather than leaving such decisions to the
regulatory process. As Waxman put it:

> To an extent unprecedented in prior environmental statutes, the pollution
> control programs of the 1990 Amendments include very detailed manda-
> tory directives to EPA, rather than more general mandates or broad grants
> of authority. . . . In addition, statutory deadlines are routinely provided to
> assure that required actions are taken in a timely fashion. More than two
> hundred rule-making actions are mandated in the first several years of the
> 1990 Amendments' implementation.[51]

The law listed 189 specific substances as air pollutants rather than
leaving it to the EPA to decide whether they should be regulated.[52]

In addition, all stationary sources of air pollution, such as coal-burning and other electric plants, would now be required to get permits from the EPA. These permits would specify how much pollution a particular source could emit and what means should be used to reduce that pollution. Moreover, a new emissions-trading system was implemented as an attempt to reduce acid rain. The EPA would specify how much sulfur dioxide a plant could emit (sulfur dioxide is converted to acid when it combines with water), but these quotas would be tradable. That is, a plant that could not reduce its emissions by the required amount could buy additional allowances from plants that were able to reduce their own emissions below their quotas. This scheme was meant to give plants that were already clean an incentive to become even cleaner, while lessening the economic disruption if dirtier plants were forced to close immediately. The total amount of allowances issued would be decreased over time so that sulfur pollution would decline.[53]

The Legacy of Clean Air Policy

Passage of the Clean Air Act Amendments of 1990 might be considered the last gasp of bipartisanship in environmental policy. By the beginning of the Democratic Clinton presidency two years later, the environment had become a partisan battleground, a condition reinforced when the Republican Party, led by Newt Gingrich, won control of House majority in the 1994 for the first time in forty years.[54] Further air pollution policy had to develop within the structure of the Clean Air Act.

Two features of this law were to prove particularly relevant for coal as climate change became a more salient issue. First, the system of tradable emissions permits developed to deal with sulfur and acid rain was generally regarded as successful and served as a model for many proposals to control greenhouse gas emissions. Second, in a development that surprised almost everyone, the procedures for evaluating additional substances as possible air pollutants turned out to be a way for bringing carbon dioxide, which was not a traditional pollutant, into the air pollution regulatory process.

CHAPTER FOUR

The Science of Coal and Climate

As discussed in the previous chapter, people have been burning coal since the Iron Age and began to worry about the resulting pollution as soon as coal use became at all widespread. During the 20th century, scientists established that the unpleasant effects of coal burning were not just smoke but also specific substances, such as particulates, sulfur dioxide, nitrogen oxides, mercury, and lead. Some of these pollutants were produced by impurities found in the coal that were either released (mercury and lead) or transformed (sulfur and nitrogen oxides) as the coal was burned. Each of these was harmful to the health of people or animals that came into contact with them, either through inhalation, through drinking water in which they had been dissolved or suspended, or in some cases through direct contact with the skin. Most of them caused other economic damage as well, such as the decay of buildings and the sterilization of lakes. Each of them could be removed from the exhaust product by technological means, although the expense of doing so was sometimes high.

However, by the late 20th century, scientists, policy makers, and the public had begun to be aware that coal and other combustible fuels also produced another kind of pollution, one that had no immediate effect on humans but might result in long-term, major deterioration of the earth's climate. Coal is composed chiefly of the element carbon, and the oxidation (i.e., burning) of coal transforms it into carbon dioxide (CO_2), the most important of the substances known as "greenhouse gases."[1] During the course of the 20th century, scientists came to realize that industrial activity had increased the amount of atmospheric CO_2 significantly, and the average temperature of the earth was rising as a result, with potentially disastrous consequences. Since this understanding is not universally accepted, it will be worthwhile to spend some time looking at the research and reasoning that led it to be accepted by almost all scientists.

The Greenhouse Effect

During the 19th century, scientists became aware that different elements and molecules absorbed and emitted light and other electromagnetic radiation at specific wavelengths. These wavelengths became known as "spectral lines" because they could be detected as either dark or bright lines in the spectrum produced when the light from a star or other light source passed through a prism. Knowledge of these lines can be used to determine the temperature and chemical composition of stars and other astronomical bodies, for example.

CO_2 is the most important of a group of gases that are transparent to visible light but absorb radiation in the infrared part of the spectrum. Since the earth receives radiation from the sun as visible light but sends it back into space as infrared radiation, atmospheric CO_2 tends to warm the planet. By the end of the 19th century, scientists had begun to wonder what effect such absorption might have on changes in the temperature of the earth over time.

The first serious analysis of this question is generally attributed to Svante Arrhenius, a Swedish chemist, in 1896.[2] Arrhenius was trying to understand how ice ages began and ended, and he decided to explore the effect of atmospheric carbon dioxide on climate as one possible mechanism. He calculated that average global temperatures would rise by 5°C if the CO_2 concentration doubled. However, he did not have data on CO_2 concentration or its change over time, so he was suggesting a possibility, not asserting that the phenomenon was occurring.

In 1901, another Swedish scientist, Nils Ekholm, was probably the first to explain heating by CO_2 and other gases by comparison to a greenhouse. English physicist John Henry Poynting coined the term "greenhouse effect," which is commonly used today.[3] Carbon dioxide and other atmospheric gases that behave in this way are called greenhouse gases; there are several such gases, but since CO_2 is the most important by far, the effect of the others is generally converted to a CO_2 equivalent. The comparison to a greenhouse is not exact; greenhouses keep heat in primarily by blocking convection, whereas greenhouse gases prevent heat loss by radiation. However, the name has stuck.

As of the mid-20th century, then, climate scientists understood the mechanism by which greenhouse gases in the atmosphere could raise the average global temperature and had some idea of the rate by which a given increase or decrease in the concentration of these gases would increase or decrease that average temperature. However, reliable data were lacking on the actual quantity and concentration of greenhouse gases and on how these quantities were changing over time.

In the late 1950s, Charles Keeling, then a young scientist at the Scripps Institute of Oceanography in California, was asked by the institute's director, Roger Revelle, to try to come up with the missing data.[4] Keeling placed recording gas analyzers in three locations: Antarctica, the slopes of Mauna Loa in Hawaii, and La Jolla, California, the location of the Scripps Institute. He also collected and analyzed flasks of air from other locations, both at ground level and from airplanes in flight. Keeling published his first results in 1960; with one short gap in the mid-1960s, the data collection has continued ever since. Each of the sampling sites was affected by local contamination—combustion of fuel at the Antarctic research station, occasional venting of carbon dioxide from cracks in the slopes of Mauna Loa, which is an active volcano, and exhaust fumes blown over La Jolla from Los Angeles depending on wind direction; Keeling was able to correct for all these. As he analyzed his data, Keeling reached two important conclusions. First, there was a regular and significant seasonal variation in the concentration of carbon dioxide everywhere in the Northern Hemisphere, but it was greater the further north one went. He found no such variation in the Southern Hemisphere. Keeling attributed this pattern to the uptake of carbon dioxide in the spring by deciduous plants, which then re-emitted it in the autumn when they dropped their leaves. Since higher latitudes in the south are mostly ocean, the hemispheric difference in seasonal variation made sense.

Second, in addition to the seasonal variation, Keeling found an increase in carbon dioxide concentration from one year to the next. He noted, "At the South Pole [where his data went back the furthest] the observed rate of increase is nearly that to be expected from the combustion of fossil fuel," or 1.4 parts per million.[5] Keeling's now familiar graphs, showing seasonal fluctuations in the concentration of CO_2 combined with a year-to-year increase, have since become known as the "Keeling curve."

Over the next two decades scientists improved their data and refined their models, leading to a better understanding of how fast CO_2 concentration was occurring and how this was affecting earth's climate. However, most of the informed public, if they had heard of the greenhouse effect at all, tended to see it as a speculative theory of something that might happen sometime in the fairly distant future. Only in the late 1980s did policy makers and the attentive public begin to see climate change as an immediate problem. Several public events helped to bring this change about.

Moving from Science to Policy

Lee, Freudenburg, and Howarth have described what they call "classical environmentalism," a model of environmental policy making that prevailed from the first Earth Day in 1970 to the inauguration of Ronald Reagan in 1981. In this model the scientific community would develop information about an environmental problem, the major environmental organizations would educate the public and put pressure on policy makers to do something about the problem, and eventually a new law would be passed or a new regulation promulgated in order to deal with the problem.[6] As was discussed in Chapter 3, this pattern began to break down with the Reagan Administration. Nevertheless, there continued to be support for environmentalism among Republican elected officials. While there was often disagreement about how best to deal with an environmental problem, most people in public life accepted the reality of those problems. The first step in pursuing a new environmental policy was therefore to bring the scientific findings into public consciousness. Two developments in 1988 played important roles in doing so: the Senate testimony of a group of climate scientists led by James Hansen and the establishment of the Intergovernmental Panel on Climate Change (IPCC). An important additional contribution to public consciousness in the United States came almost two decades later with the release of the movie *An Inconvenient Truth*.

The Importance of James Hansen

James Hansen earned his PhD at the University of Iowa studying the planet Venus, and by 1978 was at the National Aeronautics and Space Administration (NASA) in charge of an experiment to study that planet's clouds. The experiment was one of several being taken to Venus by the Pioneer mission, but before it reached its destination, Hansen had resigned from the mission in order to devote himself to the study of Earth's atmosphere instead. He was motivated to do so by the knowledge that the CO_2 level on Earth was increasing rapidly. Realizing that this change might be important for our planet's future, he turned to building a computer model of the climate.[7]

In June 1988, Senator Timothy Wirth (D-CO) convened a hearing before the Senate Committee on Energy and Natural Resources at which Hansen and several other scientists were invited to testify about climate change. Hansen presented the known evidence. First, the average global temperature was rising; at that time, four of the hottest years since records began had occurred in the 1980s. Second, greenhouse gas concentration had been increasing. And third, climate patterns were consistent with the predictions of the greenhouse effect model: temperature increases were greater at high latitudes than low, greater over continents than over oceans, and the upper atmosphere was cooling while the lower atmosphere warmed. He concluded that he was now 99 percent certain that the warming trend was caused by the greenhouse effect, not by natural variation, and that "it is time to stop waffling so much and say that the evidence is pretty strong that the greenhouse effect is here."[8] Hansen's testimony was supported by the other scientists testifying and drew supportive remarks from several senators, including Wirth. This hearing put global warming squarely on the national agenda.[9] Meanwhile, the issue was gaining a place on the global agenda as well.

The Intergovernmental Panel on Climate Change

In 1988, the year of Hansen's Senate testimony, the Intergovernmental Panel on Climate Change (IPCC) was founded under the joint sponsorship of the United Nations Environment Programme and the World Meteorological Organization. The IPCC was charged with assessing the state of scientific research on climate change, including estimates of how reliable its conclusions were, and presenting them in a comprehensive report for the use of policy makers. The panel mobilizes over 2,400 scientists who work in specialized teams to produce assessment reports; the first of these was issued in 1990, with a special supplement prepared in 1992 for the UN's Rio Conference on Environment and Development (the "Earth Summit") that year. There have now been five full assessment reports, with a sixth due in 2022.[10]

The first assessment report pointed to uncertainty about "the timing, magnitude and regional patterns of climate change," due to incomplete scientific understanding of the dynamics of clouds, oceans, and the polar ice caps, as well as of all possible sources and sinks of greenhouse gases. Nevertheless, the panel concluded that "we are certain" that there is a natural greenhouse effect that keeps earth warmer than it would be and human activities are increasing the amount of greenhouse gases in the atmosphere, which in turn would lead to "an additional warming of the

Earth's surface." They predicted an increase in global temperature of about 1°C by 2025, and 3°C by 2100.[11]

The creation of the IPCC and the publication of its first assessment report established the issue of climate change on the global agenda. The 1992 United Nations Conference on Environment and Development responded by establishing an international treaty, the United Nations Framework Convention on Climate Change (UNFCCC). The convention was signed by the United States on June 12, 1992, ratified by the Senate a few months later on October 15, and entered into force on March 21, 1994. The annual Conference of the Parties (COP) of the UNFCCC quickly became the major forum for further international discussion and action on climate.

An Inconvenient Truth

In 1983, after both Watt and Gorsuch had resigned in the midst of controversy, Reagan brought back William Ruckelshaus, the EPA's former administrator and a man highly respected by environmentalists, to head the agency again. Policy became less confrontational, and Reagan's successor, George H. W. Bush, declared himself the "environmental president." However, partisan polarization began to grow during the Clinton administration and continued subsequently. For the first time in American politics there began to be partisan disagreement about the basic scientific findings showing the existence of a problem, not only about the appropriate solution. Since the American press has a strong ethic of presenting both sides of every issue and a belief that the mass public has little patience for the complexities of scientific findings, what most people heard about climate change was that Democrats said it was happening and Republicans said it was not. Almost two decades after Hansen's Senate testimony, U.S. climate policy seemed to be at an impasse. This began to change with the release of an important documentary film, *An Inconvenient Truth*, in 2006.

The film, directed by Davis Guggenheim, focused on former Vice President Al Gore and the slide show he had developed for lectures on climate change. Gore's slides showed the reality of climate change visually with before-and-after photos of glaciers, snowcapped mountains, and the polar ice caps, along with graphs that demonstrated the clear correlation between atmospheric CO_2 concentration and temperature. The film won the Academy Award for best documentary of 2006 and was a major reason for Gore's winning the Nobel Peace Prize, which he shared with the IPCC, in 2007. While it did not end the debate over climate science, it has had a major impact on the public reception of that debate. Public support for action on climate had risen to 61 percent by 2009, although it later fell before rising again.[12]

The political struggle over climate change policy, and the place of coal in that policy, will be taken up in later chapters. But first we need to take a closer look at what is now known about the process of climate change and its likely consequences.

Data Today

The effect on climate of the concentration of CO_2 and other greenhouse gases is very firmly established. Scientists have been measuring CO_2 concentration directly since Keeling began his observations in 1958 and have been able to calculate earlier concentrations by drilling down into the Greenland ice cap; the ice traps CO_2 as it freezes, and since the lower ice was deposited first, the age of a sample can be determined from how deep in the cap it is. These data show a strong correlation with changes in mean global temperature. Of course, correlation does not prove causation. However, in this case we know the mechanism that connects the two phenomena, namely the absorption of heat radiation by CO_2. Greenhouse gas concentration is not the only factor affecting global temperature. For example, the temperature could be reduced if the atmosphere were to become less transparent to visible light, perhaps because of matter emitted by a volcanic eruption or because of the spread of smog. These are temporary effects, however, and have not prevented the global temperature rise of the past two hundred years.

The relationship between increasing CO_2 concentration in the atmosphere and the anthropogenic emission of CO_2—that is, emissions due to human activity—is also well established, but not quite as firmly so as the relationship between concentration and mean temperature. We know that anthropogenic emissions have increased and that atmospheric concentration has increased as well, but we do not yet understand the causes of changes in concentration before industrial times, and we do not completely understand all the ways CO_2 is removed from the atmosphere or at what rate such removal takes place. We do know that CO_2 emissions, CO_2 atmospheric concentration, and mean global temperature have all increased. Let us look more closely at each of these in turn.

Increasing CO_2 Emissions

It is not presently feasible to measure the total emissions of CO_2 directly. Instead a team at the Energy Information Administration (EIA) in the U.S. Department of Energy calculates annual emissions from data on the amount of coal, oil, natural gas, gasoline, and other fuels that are burned, as well as the amount of cement produced (since cement production also releases CO_2). Since we know how much CO_2 will be emitted when one ton of a given grade of coal is burned with a given degree of efficiency, it is possible to calculate how much CO_2 was emitted by the total amount of coal burned in an individual country and by adding these totals to get the total emissions for the world. This same method makes it possible to estimate the total emissions for times in the past, as long as the economic data are sufficiently complete. As of March 2017, the EIA has been able to produce estimates they consider to be reliable for every year from 1751 through 2014.[13]

The data are provided for five sources of CO_2 emissions: gas, liquids (e.g., fuel oil, gasoline, kerosene, etc.), solids (coal, charcoal, wood), cement production, and gas flaring (gas that is burned at the wellhead in oilfields as a means of disposal, rather than for fuel, which is tracked separately). In 1751, a total of three metric tons of carbon were emitted, virtually all of it from the combustion of solids. By 2014, the total was 9,855 metric tons, the largest amount from solids (4,117), closely followed by liquids (3,280). Total emissions had been 534 metric tons in 1900, 1,630 in 1950, and 6,733 in 2000. The rate of growth is rising sharply; emissions increased threefold from 1751 to 1900, threefold again from 1900 to 1950, fourfold from 1950 to 2000, and by 46 percent in the last fourteen years.[14]

Those figures represent virtually all of the human-caused, or "anthropogenic," emissions of CO_2. However, there are also emissions from natural processes such as volcanic activity and the breathing of animals. Plants both consume and emit CO_2 at varying rates, and large amounts are absorbed by the ocean. Therefore, we cannot simply assume that increased anthropogenic emissions lead to increased atmospheric concentration of CO_2; the latter must be measured separately.

Increasing Atmospheric Concentration of CO_2

As with CO_2 emissions, it is not possible to measure the atmospheric concentration of greenhouse gases at every point in the atmosphere. However, the measurement can be more direct than it is for emissions. Scientists beginning with Charles Keeling (see earlier discussion) have set up monitoring stations at a number of points to record the local concentrations of CO_2 and of some other greenhouse gases in an ongoing manner. The National Oceanic and Atmospheric Administration (NOAA) compiles estimates of the annual global concentration at sea level going back to 1980. The concentration of CO_2 in the atmosphere in 2011, as reported in the IPCC's Fifth Assessment Report, was 390 parts per million (ppm), an increase of 40 percent since 1750.[15] By 2015, it had risen to 401 ppm.[16] The concentration in 1980, the earliest year for which there was enough direct measurements to calculate a global average, was 338.8 ppm; the concentration has been rising steadily at a rate of approximately 15 ppm per decade, with slightly greater annual increases since 2010.[17]

Increasing Global Mean Temperature

As NOAA states on its climate website:

The concept of an average temperature for the entire globe may seem odd. After all, at this very moment, the highest and lowest temperatures on Earth are likely more than 100°F (55°C) apart. Temperatures vary from night to day and between seasonal extremes in the Northern and Southern Hemispheres. This means that some parts of Earth are quite cold while other parts are downright hot.[18]

However, changes in mean global temperature are a useful index of changes in the energy content of the atmosphere. Since the actual temperature is not very meaningful, data on mean global temperature are reported in degrees above or below the average temperature for a base period, a figure that is referred to as the "anomaly." If the entire 20th century is used as a base, NOAA calculates that the anomaly in mean global temperature for 2016 was 0.92°C. The last year for which the anomaly was below zero (i.e., the mean global temperature was less than the average for the century) was 1976.[19] NASA reports similar findings.

There is a lot more to climate science than these three sets of data. The team formerly led by James Hansen at NASA has developed comprehensive computer models of climate change that allow detailed adjustments and precise calculations.[20] But these data are sufficient to show that greenhouse gas emissions are increasing, that greenhouse gas concentration is increasing, and that temperature is increasing as well. Next let us look more closely at the likely consequences of these changes.

Effects of Increased Greenhouse Gases

The most striking finding of the 21st century is that climate change is having a bigger impact than expected as well as having that impact sooner than expected. The mean global temperature has risen just one degree over the average for the 20th century, which seems like very little. However, the results of the increase can be seen in the shrinking of the polar ice caps and mountain glaciers, in changes to agriculture and wild ecosystems, and in increasing frequency of severe storms. In addition, the increase in atmospheric concentration of CO_2 leads to increased CO_2 in the oceans, as well, with profound consequences for sea life.

Temperature

As mentioned earlier, global mean temperature has increased about 1°C over a 20th-century baseline. In the past, an increase of 4°C was often considered the maximum that could be tolerated without major disruption of both natural and social systems. Many now think this limit is too high and that we need to limit the increase to 2°C. This very difficult goal was adopted by the 2015 Paris Conference of the Parties to the UNFCCC, for example. But even the current 1°C increase has had major consequences for agriculture, for sea level, for the health of the natural ecosystem, and for the severity of the weather.

Agriculture and Ecosystems

The most obvious result of a global increase in temperature is that most local average temperatures will also increase. That is tautological, of course, but it is worth thinking about. Gardeners are familiar with the map of hardiness zones produced by the U.S. Department of Agriculture (USDA). Plants suited for zone six will not survive the winter in zone four and will not thrive in zone eight either. Florida lawns cannot use the varieties of grass planted in northern lawns. We can think of climate change as a poleward migration of the hardiness zones. Plants will be able to grow further from the equator than they had before, but they may not be able to migrate fast enough.

The warming climate can also expose plants and animals to new pests. In the Boston area where I live, for example, the eastern hemlock trees are threatened by an insect, the woolly adelgid, which attaches itself to the undersides of the hemlock needles and sucks the nutrients out so that the needles die. Woolly adelgid eggs are destroyed by the cold if winter temperatures drop below 20°F for at least two weeks, an event that used to be routine in Boston but is now rare. As a result, the insects have multiplied to the point where the trees may not survive.

There are also synchronization problems. Many warblers and other songbirds spend winters in the tropics but fly to the Far North to breed, eating the plentiful insects to be found there in order to get enough energy to form eggs and feed the hatchlings. The birds have evolved to use the length of the day to trigger the start of their migration so that they will arrive just as the insect hatches are beginning. However, insect cycles are driven by temperature, not sunlight; with global warming, the insects hatch earlier, while the songbirds arrive when they always did and potentially miss out on food they need to thrive.

Severe Weather

In 2005 Hurricane Katrina broke down protective levees and flooded the city of New Orleans, forcing hundreds of thousands to evacuate the city. These dramatic events captured public attention, leading many to ask whether such a heavy storm might be the result of climate change, a question that has been repeated in regard to subsequent severe storms. The answer is always both no and yes. No, any particular storm is caused by particular factors: the development of a low-pressure system over the ocean that then gains in energy as it passes over warm tropical waters, with the increased energy leading to higher wind velocity over a larger area. There were big storms well before the industrial age changed the composition of the atmosphere. But yes, as the atmosphere and the oceans have warmed, they have gained in energy so that the probability of a big storm, and therefore the number of such storms in a given year, has increased. The effect of increased energy in the atmosphere and ocean can be compared to what happens when one heats water in a pot; the water becomes turbulent long before it reaches boiling point. The ocean is not going to boil, but it is turbulent already, with local temperature differences driving both storms and currents; any increase in energy will increase the magnitude of that turbulence.

Much of the destruction wrought by Katrina was due to the failure of the levees that protected New Orleans from flooding, rather than the power of the storm itself; however, the storm also caused major damage along the coasts of Louisiana and Mississippi and was soon followed by another major storm, Hurricane Rita. All told, four Atlantic hurricanes in 2005 reached Category 5 on the Saffir-Simpson Hurricane Wind Scale, a record that still stands at the time of this writing.

The number and intensity of Atlantic hurricanes that year raised questions about the role of climate change and particularly whether this was only the beginning of a worsening trend. A study published that year concluded that there had been "a large increase in the number and proportion of hurricanes reaching categories 4 and 5" over the previous thirty-five years.[21] However, that increase has not continued in the years from 2006 to 2017. A great deal of research has been done on the relation of hurricanes to global warming, with the overall conclusion that it is too soon to tell. On the one hand, we do not have data over a long enough time period to distinguish long-term change from periodic variation. On the other hand, the predictions of theoretical models are heavily dependent on those models' assumptions.

Some researchers have found a correlation between increases in sea surface temperature and increases in the number of hurricanes in categories 4 and 5.[22] Some computer models have supported this while simultaneously predicting a decrease in the total number of hurricanes and other tropical cyclones.[23] Mallard offers a possible theoretical explanation for this. Tropical cyclones are driven by the convective transfer of heat from warm surface waters to cooler air at the top of the troposphere. However, many global climate models predict that warming will be greater in the upper troposphere than at the surface of the ocean. Since cyclone formation is driven by the temperature difference, rather than the surface temperature alone, upper tropospheric warming could make it more difficult for one to form.[24] When local variation does cause a cyclone to form, though, the extra energy available can make it more powerful.

Hurricane science is more complicated than that. There are other factors, including the amount of water vapor in the atmosphere (more moisture increases hurricanes) and the degree of vertical wind shear—that is, the variation of wind velocity with altitude (more vertical shear decreases hurricanes). The most that can be said is that no one is sure about this yet. The threat of sea level rise is much more definite and potentially very destructive.

Sea Level Rise

Increasing mean global temperature can affect sea level in two ways. Most directly, water expands as it gets warmer. This is not always the case, however; water with a temperature between the freezing point and 4°C contracts as it warms and then expands as it climbs above 4°C. Since most of the ocean is warmer than 4°C, this anomaly need not concern us. So to the extent that the warming atmosphere warms the ocean, the ocean will expand and sea level will rise as a result.

However, thermal expansion of the ocean is only a small concern; the much bigger potential problem is the possible melting of the ice caps that currently cover much of Antarctica and Greenland. Antarctica is covered by ice with an average depth of seven thousand feet. In the unlikely event that it all melted and flowed into the oceans, that increase in water volume could raise world sea level by about two hundred feet. However, Antarctica is very cold, and much of it is likely to remain so. A more immediate threat (that is, one that might develop within the next century) is the possible melting of the ice cap on Greenland, which is much smaller than Antarctica's but still with enough ice to raise the sea level by twenty feet. The Greenland ice cap is melting slowly now, enough to combine with thermal expansion to raise sea level by about twenty inches by 2100. There is also some concern that the ice cap may become unstable, with large sections breaking off and sliding into the ocean all at once, with possible catastrophic results. It is difficult to be certain about this either way.

A rise of twenty inches (and it might be more, the estimates vary) would require major expenditures in coastal cities to protect valuable property, and it would be much more serious for some small island states that would be nearly submerged. The current estimate of the IPCC is that this amount of rise probably cannot be prevented, and governments should concentrate on adapting to it. However, it remains essential to reduce greenhouse gases in order to prevent much larger possible sea level rises in the more distant future.

Ocean Acidification

In addition to climate change, atmospheric CO_2 also affects the acidity of the oceans. A significant portion of the CO_2 emitted into the atmosphere does not stay there but is absorbed into the ocean. This comprises from 25 percent to 30 percent of the CO_2 emissions from human activity.[25] The dissolved CO_2 reacts with carbonate ions to form bicarbonate ions, a process that, in turn, lowers the saturation of calcium carbonate in the seawater and increases the acidity of the ocean.[26]

Corals, shellfish, and many other marine organisms use the dissolved calcium carbonate to form skeletons or shells. As the saturation decreases, this substance becomes less available to them. The results include what has become known as "coral bleaching" and the weakening of shellfish shells. Since these animals are major food sources for other marine life, including salmon and whales, and coral reefs also provide an important habitat for young fish, the consequences can be serious.

One group of scientists have concluded that ocean pH levels are likely to drop 0.4 points by the end of the 21st century, and they note that "such dramatic changes of the CO_2 system in open-ocean surface waters have probably not occurred for more than 20 million years of Earth's history."[27]

Another group predicted a decline of 0.7 points, lower than at any time in the past 300 million years.[28] These pH changes may sound small, but since pH is a logarithmic scale, they are actually substantial. A drop of 0.3 points means that the ocean is 30 percent more acidic.

Both global warming and ocean acidification are major threats to the continued health of human civilization. The evidence for these risks is strong, but a few scientists and a larger number of political figures disagree with them. It is worthwhile to spend some time understanding why there is disagreement.

Why Some Disagree

The reasons people disagree with the need for action on climate change fall into three broad categories: self-interest, ideology, and scientific dissent. To put it simply, those who own untapped reserves of coal, oil, or natural gas will lose a lot of money if they are unable to convert those reserves into cash because of climate regulations. To the extent that self-interest influences perception, those in such a position will be less likely to agree that climate change is occurring. A larger group of people, including many conservative political activists, are opposed to government regulation and believe that government officials tend to exaggerate the need for regulation by playing up the risk of potential threats, such as climate change. This group is therefore reluctant to accept the scientific arguments that are made.

Those two groups might be dismissed as selfish in the first case, and shortsighted in the second. However, there are scientists who question the evidence of climate change on scientific grounds. This should not be surprising; science proceeds by disagreement and debate. Scientists publish their conclusions along with the data and the reasoning that led them to those conclusions, and other scientists look for flaws in the reasoning and mistakes in the data. The IPCC reports give the consensus of most climate scientists, but there have always been dissenters. It is worth examining what those dissenters have to say.

Self-Interest

It is part of the nature of capitalism that companies and industries arise, grow, and disappear at a rapid pace. Clayton Christensen, a prominent student of business, has heralded this process as "disruptive innovation," suggesting that it is both inevitable and the key to future progress.[29] Given the fluidity of capital, such disruption need not cause a crisis for the system as a whole; investors can simply sell stock in declining companies and invest in rising ones. However, some assets may be difficult or impossible to sell. If the world stops burning coal, there would be two categories of such assets: undeveloped reserves of coal (i.e., coal that is still in the ground) and coal-burning power plants. Utility companies can deal with the plants by phasing them out and replacing them with non-coal plants when they reach the end of their useful life, but anyone left holding coal reserves faces the possibility of a huge loss. The U.S. Energy Information Administration estimates that the "demonstrated reserve base" of coal in 2015 amounted to 477 billion tons.[30] The price at the end of 2015 was $31.83 per ton, a fall from the previous year.[31] At those prices, the value of the demonstrated reserve base would be over $15 trillion. The owners of those reserves thus have a powerful incentive to work against strong regulation of greenhouse gas emissions.

The coal industry made $16.1 million in political contributions during the 2012 election cycle, followed by $11.2 million in 2014 and $13.5 million in 2016. The largest amount ($266,000) went to Donald Trump's presidential campaign.[32] The industry also spent substantial amounts on lobbying against regulation of CO_2 emissions and in favor of carbon sequestration as an alternative to regulation.[33]

Over the past several years the coal industry has formed a number of different organizations for the purpose of influencing public policy. The Center for Energy and Economic Development (CEED) was founded in 1992 with members that included coal mining companies and railroads involved in coal transport. CEED supported the interests of the coal industry on such issues as climate change, mercury pollution, and regional haze. In 2009, the DeSmogBlog Project obtained a leaked copy of a 2004 memo from CEED president Steve Miller to Irl F. Engelhardt, at that time Chief Executive Officer of Peabody Energy. According to Miller, "In the climate change arena, CEED focuses on three areas: opposing government-mandated controls of greenhouse gases (GHG), opposing 'regulation by litigation,' and supporting carbon sequestration and technology as the proper vehicles for addressing any reasonable concerns about greenhouse gas concentrations in the atmosphere." Among its tactics was the mobilization of the "citizen army" of Americans for Balanced Energy Choices (ABEC) to call U.S. senators to oppose pending climate legislation, with some seven thousand calls placed; major financial support of Unions for Jobs and the Environment (UJAE), "composed of ten large industrial unions;" and "modest financial contributions" to six regional carbon sequestration partnerships established by the U.S. Department of Energy." As Miller put it, "Our belief is that, on climate change like other issues, you must be for something rather than against everything. The combination of carbon sequestration and technology is what we preach and we are looking for more members in the choir."[34]

Two years later this recognition of the need to "be for something rather than against everything" led to the consolidation of CEED and ABEC into a new organization, the American Coalition for Clean Coal Electricity (ACCCE).[35] ACCCE added electric-generating companies that use coal to the industries behind CEED and raised a budget of $45 million, triple that of CEED. The citizen army became America's Power Army and hired a lobbying firm to recruit 225,000 volunteers. The new group had two thrusts: efforts to influence public opinion, and thereby to bring pressure on members of Congress, and direct financial contributions. Executives and political action committees affiliated with the forty-eight member companies of ACCCE contributed to the campaigns of 87 percent of the members of Congress during the 2008 election cycle, a total of $15.6 million.[36]

The new group incurred serious negative publicity the next year when the staff of Rep. Tom Perriello (D-VA) discovered that six letters urging the representative to oppose the Waxman-Markey "cap and trade" bill, purportedly from the Albemarle-Charlottesville chapter of the National Association for the Advancement of Colored People (NAACP) and the Charlottesville Latino organization Creciendo Juntos, had actually been forged by an employee of Bonner & Associates, a lobbying firm working for ACCCE through a subcontract. One letter said, "We ask you to use your important position to help protect minorities and other consumers in your district from higher electricity bills." The initial discovery led to a search that found a total of fourteen forged letters, some addressed to Perriello and others to two Democratic representatives from Pennsylvania, Kathy Dahlkemper and Christopher Carney. The letters were on the letterheads of the organizations they claimed to represent but were signed by nonexistent people. Miller, the head of the ACCCE, denounced the forgery and severed the contract with Bonner & Associates.[37]

Activity based on the private financial interest of the coal industry play an important role in blocking efforts to control greenhouse gas emissions. However, that impact is magnified greatly by more generalized ideological opposition to government regulation.

Distrust of Government

The National Council for Science and the Environment (NCSE) seeks to improve the quality and accessibility of scientific information available for policy decisions and to promote the use of science by public officials. It holds a conference each January in the Washington, DC, area as one way of doing this. The 2014 conference, "Building Climate Solutions," included talks by two conservative politicians who supported strong climate action: John Gummer, Lord Deben of the UK—who at that time was the climate advisor to David Cameron, Conservative Prime Minister of Britain—and Robert Inglis, a former Republican member of Congress from South Carolina. Asked why so many conservatives rejected the strong scientific evidence that greenhouse gases were causing climate change, each man gave a very similar answer. Conservatives, they said, see liberals as people who want to raise other people's taxes and regulate other people's behavior. When they hear liberals saying that we have to do something about climate change, they think it is just another excuse for more taxation and regulation, and their distaste for those things makes them dismiss the evidence out of hand.[38]

This explanation is supported by comments submitted to the EPA in opposition to its proposed endangerment finding. For example, a letter to the EPA from Thomas A. Schatz, president of Citizens Against Government Waste (CAGW), began its comment as follows:

> CAGW is particularly concerned that if the EPA makes an endangerment finding for GHG, it will insert itself into an area that should only be the purview of elected officials, namely Congress, and not by unelected bureaucrats. EPA's involvement in regulating GHG will create a bureaucratic nightmare for businesses and other entities. It will drive up the budgets of the EPA plus state and local regulatory agencies and increase the size of these bureaucracies. The proposed regulation, if undertaken, will cost businesses and taxpayers billions of dollars. A recent study by the Heritage Foundation has estimated the cost of implementing the regulation would result in gross domestic product loses of nearly $7 trillion by 2029. This study used the least onerous assumptions for CAA regulation impacts, so the likely costs will be even higher.[39]

The letter did refer later to a belief that anthropogenic climate change was not occurring, but even this was put in the context of excessive costs:

> CAGW also believes that ongoing scientific research will ultimately prove that global warming and cooling trends are natural occurrences, that any activity to reduce CO_2 in an attempt to lower the earth's temperature will be a wasted use of tax dollars, and an unnecessary and expensive government intervention in the private markets.[40]

Similarly, a submission from the Free Market Coalition asserted, "It is now clear that setting GHG emission standards under CAA §202 would trigger a regulatory cascade throughout the Act, imposing potentially crushing burdens on regulated entities and the economy."[41]

The archetype of this view is a statement by Aaron Wildavsky, widely quoted in conservative publications, that "warming (and warming alone), through its primary antidote of withdrawing carbon from production and consumption, is capable of realizing the environmentalist's dream of an egalitarian society based on the rejection of economic growth in favor of a smaller population's eating lower on the food chain, consuming a lot less, and sharing a much lower level of resources much more equally."[42]

If this analysis is correct, then much of the opposition to greenhouse regulation is not really based on the scientific findings about climate change but is rooted in a broader conservative hostility to government action. Nevertheless, the scientific debate is important, and there is some scientific dissent from the widely accepted view that increased greenhouse gas concentration in the atmosphere has the potential to raise the mean global temperature dangerously. We now turn to this argument.

Scientific Dissent

Scientists as a group have methods to evaluate evidence and come to conclusions about theoretical claims. In the case of climate change, the IPCC has performed the most authoritative evaluation, which has affirmed that the earth's temperature is rising to an extent that threatens human health and safety. However, scientific truth cannot be determined by majority rule, and the history of science is full of cases where a small group, or even an individual, proved to be correct when the majority of scientists were wrong. It is important, therefore, to look at the actual arguments made by those scientists who question the danger of climate change.

A number of scientists have signed statements, testified, or published articles disputing the seriousness of climate change as a danger. Their most common argument is that not enough is known to justify action to limit CO_2 emissions and that such action would be costly. We shall return to the cost issue in a later chapter. As to the state of scientific knowledge, there is certainly room for improvement. In particular, as discussed earlier, the measurement of the average global temperature would be more precise if there were many more sensors, and the computer models of the climate could be improved as well. However, there is no reason to think that any error must be in a particular direction; the models are just as likely to be predicting too little temperature increase as too much.[43]

Much of the uncertainty of climate models involves feedbacks. Rising temperatures will change other things that can either raise the temperature some more or else cool it off. Some of the best-known feedbacks are the release of methane from the Arctic tundra as it thaws and the replacement of the highly reflective Arctic ice cap and various permanent snowfields and glaciers with darker surfaces as the former melt. The first would add more of a powerful greenhouse gas to the atmosphere, and the second would increase the amount of sunlight absorbed by the earth rather than reflected back into space.

Rising global temperatures would also increase the amount of water (both vapor and suspended droplets) in the atmosphere. Water is an important greenhouse gas, so an increase in its atmospheric concentration might cause further warming. However, water also helps form clouds, which have a more ambiguous effect, as they block radiation from the earth (a warming effect) while reflecting radiation from the sun back into space (a cooling effect). Whether increased cloud cover constitutes a net positive or a negative feedback depends on many details, all of which are open to dispute.

Again, much of the objection to the warming thesis related to clouds amounts to saying that we do not know enough to act. However, a more serious challenge to the theory of climate change comes from Richard Lindzen. Lindzen served as the Alfred Sloan Professor of Meteorology at the Massachusetts Institute of Technology (MIT) from 1983 to 2013 and has published many scientific articles in his field. He has made important discoveries in midatmospheric ozone photochemistry and atmospheric tides, and he was the first to explain the quasi-biennial oscillation, a pattern of stratospheric winds that gradually cycle between easterly and westerly over a period of twenty-eight to twenty-nine months. Lindzen was also one of eleven lead authors of "Physical Climate Processes and Feedbacks," chapter 7 of Working Group I's portion of the IPCC's Third Assessment Report.[44]

Since retirement, Lindzen has been appointed a senior fellow at the Cato Institute, a Washington, DC, think tank with libertarian principles, which has generally opposed regulation of greenhouse gases. Moreover, Lindzen's name was on a list of individuals and think tanks that had been funded by Peabody Energy, the world's largest privately owned traded coal company. The list was disclosed by Peabody when it filed for bankruptcy in April 2016. The amounts and dates of the payments were not disclosed.[45]

However, it would be wrong to conclude that Lindzen was paid to espouse pro-coal views in the equivalent of a bribe. It may well have been that Peabody knew of his research and thought it would be in their interest to help him do more of it. To put it differently, his being paid for the research does not necessarily mean that his conclusions are wrong. His work must be evaluated in its own right.

Lindzen began to publish articles questioning the consensus about anthropogenic climate change sometime around 1990. He is among those who have maintained that the data and the computer models commonly used to predict future warming are not sufficiently accurate to give reliable results. In 1990 he argued, "We cannot be at all sure that these small changes [i.e., in average global temperature] are not, in significant measure, due to inadequate and/or improper sampling."[46] Two decades later he asserted, "In principle, climate modeling should be closely associated with basic physical theory. In practice, it has come to consist in the almost blind use of obviously inadequate models."[47] With such arguments, Lindzen suggests that the possibility of catastrophic climate change is sufficiently doubtful that no action is justified. This position is not shared by most climate scientists; moreover, it is not clear why any error in the data or modeling might not mean that the rise in average global temperature might not be more than the models predict, rather than less. However, Lindzen has some more specific arguments as well. These are basically two.

During the 1990s, Lindzen argued that standard models of the greenhouse effect failed to take account of convection:

> One begins by recognizing that water vapor, the atmosphere's main greenhouse gas, decreases in density rapidly with both height and latitude. Surface radiative cooling in the tropics, which has the highest concentration of water vapor, is negligible. Heat from the tropical surface is carried upward by cumulus convection and poleward by the Hadley circulation[48] and planetary scale eddies to points where radiation can more efficiently transport the heat to space.[49]

This point is interesting but not very relevant to the topic at hand, since the decrease in water vapor with altitude and the poleward transport of heat by convection were both in place before the concentration of greenhouse gases began to increase. Whatever cooling effect it has will not change when CO_2 concentration increases.

In 2001 Lindzen presented a stronger argument against the climate change hypothesis, claiming that coverage of the tropical region by cirrus clouds decreased with rising surface temperature and that "this new mechanism would, in effect, constitute an adaptive infrared iris that opens and closes in order to control the Outgoing Longwave Radiation in response to changes in surface temperature in a manner similar to the way in which an eye's iris opens and closes in response to changing light levels."[50] Lindzen supported this colorfully named hypothesis by analysis of twenty-two months of data on cloudiness and temperature for the eastern part of the western Pacific gathered by the Japanese geostationary meteorological satellite GMS-5.

This remarkable feedback mechanism would mean that we would not need to worry about climate change, as the atmosphere was controlling its own temperature. However, their findings have been questioned by other scientists. In May 2001, Dennis Hartmann and Marc Michelsen of the University of Washington stated that they had reexamined the data used by Lindzen and his colleagues and the variation in cirrus cloud coverage had occurred mostly in the subtropics, over one thousand kilometers from the convective currents that supposedly caused them, so that any causative relation could be ruled out.[51] Lindzen's group published their own response a few months later. They claimed that the one thousand kilometer distance was a "red herring" and that the real issue was whether subtropical variations in cloud cover were due to convection or to the intrusion of patterns from the temperate zone.[52]

A second criticism of the Lindzen team's work was raised by Bing Lin and others at the NASA Langley Research Center. Lin's team used direct satellite observation of longwave radiation to replace the values Lindzen's group had derived from theoretical modeling and found that the results changed.

> The modeled radiative fluxes of Lindzen et al. are replaced by the Clouds and the Earth's Radiant Energy System (CERES) directly observed broadband radiation fields. The observations show that the clouds have much higher albedos and moderately larger longwave fluxes than those assumed by Lindzen et al. As a result, decreases in these clouds would cause a significant but weak positive feedback to the climate system, instead of providing a strong negative feedback.[53]

Chou, Lindzen, and Hou replied that their method of determining albedo and longwave radiation was the correct one. They had "specified subjectively the OLR and albedo . . . while requiring that the mean OLR and albedo of the Tropics be consistent with the Earth Radiation Bdget Experiment inferred values." Lin's group, in contrast, had derived these values by interpreting the satellite photographs. Chou et al. believed that Lin's interpretation had been faulty and that their own subjective specification was more accurate.[54]

Clearly this issue has not been settled. There may or may not be an iris effect, and if there is it may or may not be large enough to eliminate the warming that would otherwise take place. We do know that warming of 1°C has occurred already, despite any iris effect that may be operating. Given the uncertainty, and the magnitude of the threat to society, it would seem prudent to seek to lower greenhouse gas emissions. In the next chapter, we look at some of the options for doing so.

CHAPTER FIVE

Can We Live with Coal? Can We Live without It?

At this point, most scientists and most governments have agreed that CO_2 emissions constitute a serious threat to the environment.[1] Moreover, the consequences of this threat may be extremely severe. On the other hand, society depends on the energy derived from coal (and other carbon-based fuels) to produce the goods and services that we have come to value. In 2016, the world consumed 43.7 petawatt hours (PWh) of energy from coal.[2] In order for CO_2 emissions from coal to be eliminated, alternative sources for this amount of energy would have to be found. In this chapter, we shall examine various proposals for obtaining this energy from sources that will not increase atmospheric concentration of GHGs. Many of these proposals are feasible technologically. However, at this time none is being implemented fast enough to keep the stock of GHGs in the atmosphere below critical levels. This is primarily due to considerations of cost. If the price of coal (and oil) included the costs of the environmental damage it causes, movement away from GHG emissions could be achieved without major difficulty. As things stand, however, little movement is taking place.

Proposals to counter climate change can be grouped into several broad categories: developing technology to burn coal without releasing CO_2 into the atmosphere (sometimes referred to as "clean coal"); "geoengineering," a term used for proposed large scale projects to either reduce the inflow of solar energy to the earth or to remove CO_2 from the atmosphere; and renewable energy solutions to replace coal and other carbon-based fuels with other energy sources while reducing the total energy needed by society. Let us begin with clean coal.

Clean Coal

The phrase "clean coal" was first used during the 1980s during the debate about acid rain, as discussed in chapter 3. It referred to coal with low sulfur content; reductions in sulfur emissions could be achieved either by using costly scrubbers to remove the sulfur from the smoke or by burning low-sulfur, or "clean," coal.[3] However, while coal can be clean of sulfur, it cannot be clean of carbon; the carbon is what gets burned to obtain energy. As used today, the term clean coal does not refer to the coal itself but to various technologies to capture and store the CO_2 after the coal is burned. A more accurate name is carbon capture and storage, or CCS. The term "carbon sequestration" is sometimes used for this as well. There are several methods for capturing CO_2, some small plants have been built as pilots, and two larger plants have begun to operate. However, it is increasingly doubtful whether any of these processes would be commercially viable soon enough and on a large enough scale to be useful.

Whatever its actual effectiveness, clean coal technology is popular politically. We saw in chapter 4 that the coal industry supports it because "on climate change like other issues, you must be for something rather than against everything."[4] Whether or not the technology can be made to work, it allows the industry to demonstrate that it is trying to deal with the problem of greenhouse gas emissions. On the other hand, some elected officials who support action to limit greenhouse gas emissions support research into clean coal technology as a way of showing that they are not anti-coal. For example, President Obama announced a national clean coal strategy in 2010. The presidential memo defining the strategy began:

> For decades, the coal industry has supported quality high-paying jobs for American workers, and coal has provided an important domestic source of reliable, affordable energy. At the same time, coal-fired power plants are the largest contributor to U.S. greenhouse gas emissions and coal accounts for 40 percent of global emissions. Charting a path toward clean coal is essential to achieving my Administration's goals of providing clean energy, supporting American jobs, and reducing emissions of carbon pollution. Rapid commercial development and deployment of clean coal technologies, particularly carbon capture and storage (CCS), will help position the United States as a leader in the global clean energy race.[5]

Coal mining is economically important in Ohio and Pennsylvania, each considered a swing state in presidential elections, and embracing clean coal served as an attempt to win more votes in those states.

Given these political considerations, one might question the sincerity of those who advocate carbon sequestration and storage. Private industry has certainly not committed enough resources to developing CCS as a solution.[6] However, many believe that CCS must be part of any effort that succeeds in limiting global temperature increase to 2°C. Before it can be deployed effectively, several problems must be solved.

An effective carbon sequestration technology must confront three basic challenges. First, it must capture the CO_2 from the exhaust gases without decreasing the energy output too much. Once captured, the CO_2 must be stored safely for a very long period of time—in principle, forever. In addition, the captured gas must be transported from the plant where it is produced to the storage site, sometimes hundreds of miles away. Moreover, these technical problems must be solved at a cost that preserves coal's commercial viability as a fuel. Several solutions to these problems have been proposed and some have been developed to the point of operating pilot plants. However, as of 2018, only two large-scale CCS plants are operating, and each of these has serious problems.

At present the most likely technology for CO_2 capture appears to be scrubbing the exhaust stream with amines (derivatives of ammonia, NH_3, in which one or more of the hydrogen atoms has been replaced with something else). The flue gas is cooled and mixed with amines, which absorb the CO_2; the resulting product is then separated out and heated, breaking the chemical bonds and leaving a mixture of CO_2 and water; finally, the water is condensed out, leaving pure CO. Amine scrubbing is well understood, has been used since 1930 to remove CO_2 from natural gas, and has the great advantage from the utility industry's point of view that it can be added to existing plants.[7]

One problem with amine scrubbing is that it reduces the energy output of the plant, since energy is consumed in the process of scrubbing the flue gas and then compressing the resulting CO_2. Currently the reduction is about 30 percent of the power output; however, there are estimates that this can be reduced to about 20 percent as the technology matures.[8] This energy must be replaced by production elsewhere.

Once the CO_2 has been removed from the combustion products, something has to be done with it. Breaking it up into pure carbon and oxygen would be possible, but doing so would require all the energy that was obtained by burning the carbon in the first place; in fact, it would require more, since some energy is always lost to friction, leakage, or other forms of waste.[9] Since the point of removing the CO_2 is to keep it out of the atmosphere, it cannot just be released but must somehow be stored. Moreover, it must be stored for a very long time, effectively forever, since releasing it in one hundred or one thousand years would only postpone the warming effect. A number of storage methods have been proposed. The most feasible are storing the gas in cavities deep under the earth or converting it chemically into a mineral that is solid and stable.

Carbon dioxide can be pumped deep underground where it can be stored in the pores of the rock. This will not work if something else is currently occupying those pores, so the best alternative is to insert carbon dioxide into the vacant space left after petroleum has been extracted. In some cases, the pressure of the CO_2 can be used to extract more petroleum, a process known as enhanced oil recovery (EOR). (This can be regarded as either killing two birds with one stone or defeating the purpose of carbon sequestration by making more carbon available to burn.) Whether the CO_2 is used for EOR or simply stored, the site must then be monitored for leaks for a very long time.

The problem of leaks can be avoided by a different storage method, pumping the gas into rocks that will react with it to form a stable mineral. There are at least two ongoing attempts to test the feasibility of this method. In Iceland the CarbFix project pumps CO_2 mixed with water about fifteen hundred feet down into a volcanic basalt formation; the basalt reacts with the water and CO_2 to form calcites. Injecting water along with the CO_2 adds to the expense considerably; however, the cost of monitoring is avoided, so the method may prove to be cheaper than that of storing gas once it is developed further. A somewhat different test is underway at a site near Wallula, Washington, in the United States. In this case, only CO_2 is injected into a basalt formation that is capped so that the gas is held in the rock. It then reacts with the basalt, but more slowly than if water had been added; the scientific team running the project expects mineral formation to be a matter of years. Further analysis and development will be required to see if the process works in the field and whether the cost makes it competitive with long-term storage.[10]

The technology for underground storage is more developed. The first full-scale carbon sequestration project at a power plant, Boundary Dam Carbon Capture and Storage, opened in October 2014 as part of an existing power plant in Saskatchewan. The project replaced an old coal-fired boiler with a new, more efficient one and added an amine scrubber. Not only the CO_2 but also sulfuric acid and fly ash were sold commercially to help cover the cost of the plant. The CO_2 was transmitted by pipe for forty miles to the Weyburn oil field, also in Saskatchewan, to be used for EOR.[11] Initial projections were that the plant would remove 90 percent of the CO_2 from the exhaust, the equivalent of taking 250,000 cars out of operation for that period. However, performance lagged far behind that during the first eighteen months of operation. Particles of ash got through filters and contaminated the amines in the scrubber, reducing its efficiency, and considerable amounts of CO_2 were lost through leaks, so that only about 45 percent of the CO_2 was captured. There were several cost overruns and unexpected repair expenses; these combined with expected costs (for example, the scrubbing process requires about 20 percent of the plant's generating power, so less electricity is produced) to raise the cost of electricity by 100 percent. SaskPower, the utility that operates the plant, argued that such problems were inevitable in a new technology and that they were learning valuable lessons for the construction of better plants in the future.[12] Critics countered that greater GHG reductions could have been obtained by spending the same amount of money on getting energy from wind, an abundant resource in Saskatchewan. One report calculated that it would have cost $1 billion Canadian less to generate the same amount of electricity from new wind plants. The real purpose of the project, these critics claim, was to enable SaskPower to save its sunk costs in the Boundary Dam plant, which would have had to close in 2020, and to allow for the construction of new plants to take advantage of its cheap nearby coal deposits; Canadian law prohibits construction of coal power plants after July 1, 2015, unless they are fitted with CCS technology.[13]

A second power plant with CCS, Petra Nova in Texas, opened on December 29, 2016. The carbon capture apparatus was retrofitted to unit 8 of the WA Parish power plant near Houston. As with the Boundary Dam project, the captured CO_2—in this case, 90 percent of it—is used in EOR in a nearby oil field that is partly owned by the owners of the power plant. The company predicts that the plant will pay for itself in ten years.[14] Petra Nova and Boundary Dam are the only commercial-scale power plants with CCS currently operating. Other CCS operations mostly involve the addition of storage to CO_2 that was being produced already, either as a byproduct (for example, of natural gas production) or expressly for the purpose of EOR.[15]

A third proposed CCS power plant, the Kemper County energy facility in Mississippi, was planned as the largest such plant in the world. The plant was intended to convert lignite coal into synthetic gas and then remove carbon dioxide from the combustion products by amine scrubbing. The CO_2 would then have been used for EOR. The planners hoped to reduce CO_2 output by 65 percent. As of 2017 the project was five years late and had spent $7 billion, $4 billion over its budget (including subsidies of $382 million from the U.S. Department of Energy and $800 million from rate-payers), and the developers—Southern Company and its subsidiary Mississippi Power—requested a further rate increase from the Mississippi Public Service Commission. The commission denied this request, recommending that instead the plant be switched to natural gas as fuel. The two companies announced on June 28, 2017, that they would make this change and cease developing the CCS technology for the plant.[16]

The Kemper project had a history of mismanagement and possible fraud, for which disgruntled stockholders are now suing the operators. In 2014 an engineer on the project, Brett Wingo, decided to become a whistleblower, saying that he felt "a duty to act." Wingo turned over internal documents and recordings of conversations that he had made both to officials of Southern Company and Mississippi Power and to the Mississippi Public Service Commission; he later provided them to the *New York Times* as well. These materials included statements by responsible officials that the delays and cost overruns were due to mismanagement and possibly fraud.[17] Further investigation by the *Guardian* over the next year found strong evidence that company officials had concealed the cost overruns for years while telling regulators and investors that all was well with the project.[18]

The third problem in CCS is how to transport the CO_2 to a storage facility. This is best done by pipeline, and since there are a number of CO_2 pipelines operating currently, this is less a technical problem than simply an additional expense.[19]

It would appear that CCS is technically feasible. However, its development lags far behind what is needed to keep the increase in the mean global temperature at a safe level. Of the first three full-sized CCS power plant projects, one has been canceled and one is seriously underperforming. Assuming that there are no unexpected problems during the Petra Nova project's initial period of operation, it is a promising development. However, the pace of CCS development is too slow. In an analysis published in 2009, the geoscientist Stuart Haszeldine, who favors the development of CCS, argued that "to change black fuel into green energy, the acceleration and scale-up of CCS is required, from tens of power plants within five years, to hundreds of large plants by 2025, and then to thousands of small power plants by 2035."[20] At around the same time the International Energy Agency published its "CCS Roadmap," an infographic summarizing the organization's analysis. It states, "Without CCS, overall costs to halve emissions by 2050 rise by 70%. This road map envisions 100 projects globally by 2020 and over 3,000 projects in 2050."[21] Actual development lags far behind this goal.

The basic reason for the lag is financial. Most plans for CCS assumed that there would be either a tax on GHG emissions (often called a "carbon tax") or a cap-and-trade system that would compel utilities to buy emissions allowances if they exceeded their quotas. Neither of these has been enacted in the United States or China, and in the EU, policy miscalculations led to a surplus of emissions permits, which consequently lost much of their value. Without them, CCS projects require either a government subsidy or an opportunity to sell the CO_2 for use in EOR at a high enough price; the Petra Nova project had both of these advantages.

We can conclude that CCS could be a part of the solution but only if billions of dollars more are put into its development. The required funds could take the form of direct government subsidies, a tax on carbon emissions that would make it more profitable to reduce those emissions, a cap-and-trade regime that places a value on emissions reductions, or some combination of these methods. The obvious question, then, is whether CCS would be the best use of this money or whether it could better be spent developing sources of energy that do not produce greenhouse gases. Before turning to this question, however, let us take a brief look at proposals to control climate change by technological interventions on a planetary scale. These proposals are known collectively as geoengineering.

Geoengineering

The term "geoengineering" refers to efforts to reverse global warming directly through the use of technology. Geoengineering projects fall into two broad types: carbon dioxide removal (CDR) and solar radiation management (SRM). CDR takes greenhouse gases out of the atmosphere, thereby restoring the planetary balance of energy absorption and radiation. SRM, on the other hand, seeks to reduce the amount of energy reaching the earth from the sun, so that the balance will be retained despite a lower rate of outward radiation.

CDR aims to take CO_2 out of the atmosphere directly. If successful this would decrease GHG concentration faster than by the decarbonization of energy production. For example, we could build large amine scrubbers, similar to those used in the Boundary Dam project, but feed them with air from the atmosphere rather than the exhaust from a power plant. Carbon dioxide in the atmosphere is much more thinly concentrated, so we would need very large scrubbers, and very many of them, at a very high expense, but the process is technologically feasible. We would also have to develop better storage solutions, but this too could be done with enough time and money. However, the magnitude of each that would be required is daunting.

Another proposal is to increase the growth of oceanic algae, which would remove CO_2 from the atmosphere. If the algae then died and fell to the bottom of the ocean, GHG concentration would decrease as a result. In much of the ocean, algae growth is limited primarily by a shortage of iron, a necessary component of chlorophyll. In preliminary experiments, sprinkling an area with pulverized iron did produce the expected algae growth; however, the dying algae floated to the surface rather than sinking, so that the CO_2 they had consumed was returned to the atmosphere rather than sequestered. Other experiments have been more successful, but there are doubts about whether the amount of CO_2 that could be sequestered in this way would be sufficient as well as about the effect the accompanying withdrawal of nutrients would have on life in the ocean.[22]

A final CDR option is known as biological energy carbon capture and storage (BECCS), the production of energy from biomass combined with existing CCS technology. If the biomass used is grown for the purpose (rather than obtained by cutting existing forests, for example), BECCS would produce energy while decreasing overall GHG concentration, so it is an attractive idea. However, it has its limits. In order to sequester the CO_2 that would be emitted if all the world's coal reserves were burned, a very large area of land would be required, land that would not be available for food production. And, as with CCS from coal, the amine scrubbers required would be massively expensive and storage would continue to be a problem. To the extent that BECCS was able to replace coal (and oil and gas) consumption, it could be part of the solution, but it would be an expensive part.

Solar radiation management involves blocking some of the sun's radiation before it reaches the surface of the earth. Various people have proposed a number of ways in which this might be done, ranging from the far-fetched to the relatively straightforward.

At the far-fetched end of the spectrum, we might put things into orbit that would reflect back some of the solar radiation before it reached the earth. Possibilities include truly gigantic sheets of very thin, transparent plastic film or very large numbers of reflective bits of metal or other shiny objects. The former is purely a concept, at least at this time; the material for the films has not been developed, and no one has a very good idea of how to keep such a film in place. (The center of gravity of a sheet could be put into geosynchronous orbit, like a communications satellite, but gravity would pull unevenly at the edges so that it might rip, bunch up, or be moved out of place.) Material does exist for the latter, but the task of getting it into space would require a massive increase in the construction and launching of rockets, which in turn would emit further GHGs into the atmosphere.[23]

Sulfate aerosol injection is a more feasible proposal. It involves putting small particles of sulfur oxides into the upper atmosphere so as to make the atmosphere less transparent to visible light. Essentially, this would mean creating stratospheric smog. The aerosols could be released from aircraft as part of the normal routine of long-distance flights.[24] To this date no such projects have been approved because of concerns over possible negative effects of the sulfates, which might catalyze the formation of ozone-depleting chemicals or descend to pollute the lower atmosphere, with deleterious effects on human health.

A final SRM proposal is cloud brightening, the injection of small droplets of water into the atmosphere in order to make clouds reflect more sunlight. This proposal is at a preliminary stage as not enough is understood about the needed size of the droplets and how they could be injected successfully.[25]

A general problem with all varieties of SRM is that any projects that blocked some of the solar radiation coming to earth would affect plant growth, including agriculture, with possibly deleterious effects on the supply of food, and might also decrease rainfall.[26] Solar radiation management also poses a risk of catastrophic failure; since GHG concentration would continue to increase, if the intervention ever failed there might be a rapid rise in temperature of several degrees.[27]

Because of these possible negative consequences, and the huge costs that would be involved, none of the SRM proposals seems likely to move to the development stage at any time in the near future; however, they may be reconsidered if other means of limiting climate change should fail. The advantage of alternative energy sources, such as wind, solar, hydroelectric, tidal, and geothermal, is that they do not produce CO_2 at all, thus the expenses of carbon capture and storage are avoided. In the next section, we shall consider whether these alternative sources can be developed fast enough.

Green Solutions

If modern society is to survive without burning fossil fuels, energy has to be produced in other ways. These can be classified into three categories: renewable energy, nuclear power, and conservation. Energy sources are renewable if they are reproduced by natural processes, almost always driven ultimately by the energy earth gets from the sun, either directly (solar power) or because energy from the sun moves water from the ocean to the headwaters of rivers (hydroelectric energy), puts masses of air into motion (wind), or causes plants to grow that can then be burned (biofuels). Tidal energy comes from the gravitational force of the moon, instead. Geothermal energy, which is produced by tapping the earth's inner heat, is not truly renewable, but it is generally regarded as such because the stored energy is so much greater than our capacity to use it.

Nuclear power, based on fission or fusion, is not renewable. It produces energy from the fission of radioactive heavy metals, of which there is only a finite supply. However, it does not produce greenhouse gases, so it could serve as a temporary replacement for fossil fuels if the other problems associated with it could be solved.

Finally, energy conservation is not a source of energy at all, but it could make a large contribution to solving the problem. It cannot be the whole solution since we still need energy, but increases in energy efficiency could make the amount we need considerably less. Let us begin by considering renewables.

Renewables

Nothing is truly renewable in any absolute sense. In the long run, the sun will explode in a fiery nova, destroying the earth in the process. In the still longer run, the universe will move closer and closer to a state of absolute entropy where everything is cold and dead. However, as long as the sun still shines and the moon continues to revolve around the earth, it is possible to tap the energy these phenomena produce for human use. We begin with the most direct of the renewable options, solar energy.

Solar

The earth receives a huge amount of energy from the sun. About 174 petawatts (PW) reach the upper atmosphere. About 30 percent of this is reflected back into space, and the atmosphere absorbs a certain amount, leaving about 87 PW at the earth's surface. This comes to about three or four exajoules of energy from the sun every year. This is roughly eight thousand times the total world annual consumption of fossil fuels as of the 1990s.[28] Put differently, the sun gives us enough energy every hour to power modern society for one year.[29]

Much of this energy goes to heat the oceans and the landmasses of the earth, as well as powering winds and ocean currents; however, a great deal remains for potential human use.

Direct forms of solar energy are often divided into *passive* and *active* forms. Passive solar uses the heat of the sun directly and can be maximized by design, for example by building structures with south-facing windows and heat sinks, such as blocks of stone or tanks of water, that will absorb energy while the sun is shining and emit it after dark. We shall return to passive solar when we consider energy conservation later in this chapter.

Active solar involves using solar radiation to heat water or generate electricity; in the latter case, the electricity produced can then be used in turn to do work of various kinds. The generation can be direct, through photovoltaic panels, or solar radiation can be concentrated to boil water, which can then drive the turbine of a conventional power plant, in essence simply replacing coal, gas, or oil as a fuel. Photovoltaic panels can be concentrated in large solar farms, thereby acting in ways similar to a conventional power plant; however, they can also be dispersed to sites where the electricity will be used, ranging from a single panel powering a road sign to solar roofs on buildings. Such dispersed siting is known as "distributed solar." It can be more efficient than concentrated plants since transmission costs are lower, but it can be difficult to integrate distributed solar production into the existing electric grid.

One of the challenges faced by solar electricity is that it is produced only when the sun is shining. Electricity must be consumed at the time it is produced. At times of high demand, or when a plant has to be shut down, backup generators are activated to produce more electricity; if this fails to meet the demand, blackouts will result, and may possibly trigger a system shutdown. Currently solar energy is only a small part of the total electricity generated and relies on the same fossil fuel–fired backup generators as the rest of the system. However, if fossil fuels are ever to be eliminated from the mix, some means will have to be found to store energy during times of high production, such as sunny days, to be used at night or in cloudy conditions. The simplest method would use batteries that could be charged in the daytime and used at night. Big enough batteries do not yet exist, but research is ongoing; the batteries would not have to be large enough to run the whole electric system from a central point but could be installed in buildings, replacing size with numbers. Energy can also be stored in other ways, such as pumping water uphill to a reservoir during sunny hours and letting it flow back through a turbine when electricity is needed, but these methods have their own environmental problems.

Solar electricity production is growing very rapidly. For the United States, the Energy Information Administration calculates that solar electricity from utility-scale plants (defined as one megawatt or more in capacity) grew from 864 petawatt hours in 2008 to 52.957 petawatt hours in 2017. The agency did not attempt to estimate electricity from distributed photovoltaics until 2014, when the total was 11.233 petawatt hours; this has grown to 24.139 petawatt hours in 2017.[30] Solar power accounts for only about 1.3 percent of the electricity produced in the United States, however.[31] On the global scale, 3.65 PWh of electricity was produced in 2016, compared to 43.6 PWh from coal.[32] Although solar energy's share is growing exponentially, it must be scaled up considerably if solar is to replace coal. However, solar power need not replace coal entirely; it will be joined by other renewable energy sources, one of which is waterpower.

Hydroelectric Power

Solar radiation not only warms our planet directly and drives solar electric production, but it also evaporates very large amounts of water from the rivers, lakes, and oceans of the earth. The resulting vapor rises into the atmosphere, condenses into clouds that are moved about by the winds, and then falls back to the earth as rain and snow. Most of the rain falls directly back into the oceans, since these cover the majority of the earth's surface, but a significant portion falls on higher altitude areas. Any water that is precipitated at a higher altitude than it was evaporated from has gained in potential energy, which is released as the water flows downhill to the sea. This energy can then be tapped to do work.

Perhaps the simplest use of waterpower is for downstream transport. Boats or other objects are placed in the water to float downstream. Such transportation works only in the downstream direction but has been useful historically for moving products from remote agricultural or lumbering areas to downstream processing plants and ports. In the vast Mississippi River watershed, flatboats were the dominant mode of shipping from their first use in 1782 to the time of the Civil War. Initially farmers themselves would build boats and hire boatmen to float their crops downstream to New Orleans, where the crops would be sold and the boats broken up for timber. With the coming of the steamboat, the boatmen could return home more quickly, enabling them to make several voyages each year; eventually, however, steamboats were to displace flatboats completely.[33] In the old growth forests of northern New England, no boats were needed; the logs themselves were floated downstream, at first to be shipped across the Atlantic to be used as masts for the British Navy and later to downstream sawmills in such cities as Bangor, Maine. These river log drives continued until the old trees had all been cut; the last drive was in 1970.[34]

Technological developments to make more efficient use of waterpower can be traced back as early as the 4th century BCE in India, China, and the Roman Empire. Early examples include the use of water wheels to lift water into irrigation ditches or to turn millstones for grinding grain into flour.[35] In the early Industrial Revolution, factories were built along the banks of rivers with machinery driven by complex systems of belts connected to water wheels.[36] As technology developed further, waterpower was replaced by steam engines and then by electricity. However, while waterpower is no longer much used to drive machinery directly, it is an important source of electricity. As of 2015, hydropower plants produced around 4 PWh, 16.6 percent of the world's electricity, and were by far the most important renewable electricity source.[37]

Hydropower has considerable potential for growth. The International Energy Agency calculates the technical potential of hydropower (defined as "the annual energy potential of all natural water flows that can be exploited within the limits of current technology) as 16.4 PWh per year, or more than four times the amount produced at present. Perhaps a more realistic prospect is that hydroelectric production could be raised to 6 PWh/year by 2050.[38] For comparison, coal-fired plants produced 9.282 PWh in 2016; hydropower could go a long way toward replacing coal.[39]

However, hydropower is not entirely good for the environment. The preponderance of hydroelectric plants use dams to collect the water and regulate its flow, and these dams are often destructive in their own right. Since they flood desirable riverfront land, dams remove land from food production and displace those who live along the river. The World Commission on Dams (WCD), which had been established jointly by the World Bank and the International Union for Conservation of Nature (IUCN), estimated that dams have displaced between 40 million and 80 million people.[40] Proposed dams in the Amazon and in the rivers leading to James Bay in Canada threaten the viability of indigenous cultures. And while dams replace fossil fuels, they can generate significant methane emissions unless the flooded land is first cleared of all vegetation. The WCD developed a set of guidelines to assure that dams are environmentally and socially beneficial; however, there is no clear means of enforcing these guidelines.[41]

Despite the controversy, new large dams continue to be built and with very large hydroelectric power plants attached. These include the 22.4 MW Three Gorges Dam in China, which began operations in 2008,[42] and the 14 GW plant at the Itaipu Dam on the border between Paraguay and

Brazil, which produced 0.103 PWh, the largest amount ever generated by a dam, in 2016.[43] The proposed 11.2 MW Belo Monte Dam in Brazil has been delayed by massive organized opposition, but it continues to be pursued by its developers and was licensed by the government in 2011. Large dams often face opposition due to the people whom they will displace, their effect on river ecology, and other environmental challenges. For this reason it is difficult to predict the extent to which society will be able to replace energy from coal with that from hydroelectric plants.

Wind Power

Wind power, like waterpower, has been in use for a very long time. Early uses included the propulsion of sailboats and the transmission of power through gears and belts to turn mills for grinding grain or to pump water from wells. The first use of wind power to generate electricity was in 1887 in separate experiments by James Blyth in Scotland and Charles Brush in the United States.[44] Wind power has the advantage that it requires neither a source of fuel nor a streamside location; its disadvantage is that the power from a given wind-driven turbine is available only when the wind is blowing at that turbine's location. As a result, early use developed in remote locations when constant operation was not required. Examples of such uses include charging batteries or pumping water from a well into a tank for use later.

Modern electric distribution grids have the capacity to shift rapidly from one source of power to another. Thus inexpensive wind power can be used when it is available and replaced by backup generators located elsewhere when it is not. The rise in the cost of oil associated with the OPEC embargo of 1973 stimulated increased construction of wind power plants, an increase that has continued. As of 2015, total global wind power capacity was equal to 433 GW.[45] Two years later, this capacity had increased by 25 percent, to 540 GW.[46]

Large wind power projects can also face environmental opposition due to health, aesthetic, and ecological considerations.[47] Wind turbines make noise, and this noise may have adverse effects on those who live nearby. They can also disrupt what would otherwise be a beautiful scene of pristine nature. Since wind turbines are usually mounted high above ground, they pose a hazard to birds that collide with them, many of which are killed as a result. All of these concerns can be lessened by careful design and siting of the wind turbines; however, the need to do so puts limits on the degree to which wind energy can be increased. In particular, with presently available technology, wind turbines, unlike solar panels, are probably not suitable for distribution to the household level. On the other hand, they are more suitable than solar panels for use in agricultural contexts since they do not block a great deal of sunlight.

A final renewable fuel is nonfossilized biomass, plants grown at least in part in order to be burned for energy. Since bioenergy does produce GHG emissions, it is somewhat different from other renewables in its effect on climate. We shall examine this next, before turning to some alternative energy sources that are not renewable.

Biofuels

Wood is humanity's oldest fuel and is still widely used for cooking and heating in less developed parts of the world. In arid regions the use of firewood has led to deforestation, but in regions with more rainfall, forests can be managed to produce wood on a sustainable basis through practices that have been known for over three centuries.[48] However, the combustion of wood produces not only GHGs but many other pollutants as well, so wood is not a suitable replacement for coal in general use. Two other categories of bioenergy have more potential: the production of ethanol to be mixed with gasoline in motor fuels and the fermentation of organic waste to produce methane that can then be burned.

The making of beer, wine, whiskey, vodka, and other alcoholic beverages all involve the fermentation of ethanol from plant sugars, with or without distillation to increase the alcohol concentration. By eliminating those techniques that are meant to assure potability and improve taste, pure ethyl alcohol (ethanol) can be produced at a relatively low cost. The use of such alcohol as a motor fuel has been advocated by the corn industry for close to one hundred years and has been supported by public policy since 1980.[49] Today ethanol is the major beneficiary of the Renewable Fuel Standard (RFS), which requires that gasoline contain a periodically increasing amount of fuel made from renewable sources.

Ethanol is one of those cases where the economic interests behind a policy are more important than that policy's stated goals. Ethanol was advocated in the 1920s as a way to provide against the possibility of an oil shortage, in the 1990s as an oxygenate to make gasoline burn more cleanly, and today as a way to reduce GHGs. Regardless of these avowed purposes, the RFS increases the demand for corn and supports the industry that has grown up to process it.[50]

Beginning with the Energy Policy Act of 2005, inclusion of renewable fuels in motor vehicle fuels became a federal requirement. Renewable fuels are defined as those with life-cycle GHG emissions lower than petroleum based fuel. Life-cycle emissions are the difference between the CO_2 removed from the atmosphere when the plants are grown and that emitted during the production and combustion of the fuel derived from those plants. There are separate standards for "conventional biofuel," derived from the starch in such plants as corn, which must lower life-cycle emissions by 20 percent to qualify, and for "advanced biofuels," which include fuels derived from cellulose, sugar, vegetable oil, and waste grease and must show a life-cycle GHG reduction of 50–60 percent. Inclusion of renewable fuels is required in annually increasing amounts, reaching 36 million gallons by 2022.[51]

Most environmentalists oppose the RFS for a number of related reasons. Its production removes agricultural land from food production and involves heavy use of the chemical fertilizers that then enter the rivers and contribute to the creation of dead zones in the Gulf of Mexico and elsewhere. Many feel that this is too high a price to pay for a technology that does not actually reduce greenhouse gas emissions but at best only holds them constant. Since existing automobiles cannot run on pure ethanol, the latter must be mixed with gasoline; it can never become a zero-net-carbon fuel. The political power of the ethanol industry has kept the RFS alive, but it will probably not be part of the long-term solution to GHG emissions.

Solar power, waterpower, wind power, and biofuels are the major sources of renewable energy. Geothermal and tidal power, which I have not discussed here, are locally important in some places but not likely to make a major contribution to global energy needs. However, before we assess renewable energy's capacity for replacing coal, we need to look at two other options: nuclear power, which is not renewable but does not produce GHGs, and energy conservation, which is not a source of energy but can make the energy we have go much further. Let us turn next to nuclear power.

Nuclear Power

The nucleus of an atom consists of a combination of positively charged protons and uncharged neutrons. Since like electrical charges repel each other, there must also be some binding energy to hold them together.[52] Under the right conditions, a nucleus may split into one or more smaller nuclei, a process called "fission." If the initial nucleus is large and radioactive, the smaller nuclei require less binding energy than their parent, and the excess energy is available for use. Fission occurs spontaneously in nature, and it can be sped up, either to make a nuclear bomb or, if the speed of fission can be controlled, to heat water or some other liquid to drive an electric generator.

Nuclear fusion is the reverse of fission: two small nuclei are forced to combine to form the nucleus of a heavier element. Typically, nuclei of deuterium, a form of hydrogen with an extra neutron, are combined to form helium. Fusion produces much more energy than fission; however, it also requires a great deal of energy to force the initial nuclei to combine. If the purpose of the fusion is a large explosion, this can be and has been done. However, to use fusion as an energy source requires that the reaction be controlled. To date, all attempts to create such fusion either consume more energy than they produce or have not been capable of being maintained for more than a fraction of a second. This problem may be solved sometime in the future, but this is not likely to happen soon enough to help replace coal as a source of energy.

Nuclear fission, on the other hand, has long been used to power both commercial electric generating plants and ships. Most environmentalists have opposed such development, primarily because of the disastrous consequences of accidents when they occur and because no good system exists for storage of the highly radioactive waste that is produced by the plants.[53] More recently, some environmental writers have argued that nuclear power may be the only feasible way to eliminate CO_2 fast enough to prevent disastrous climate change.[54]

It would require another book to examine this argument. Whatever its merits, however, to assert that it would be "feasible" for nuclear power to replace coal ignores the history of the last three decades. Disastrous accidents at Three Mile Island, Chernobyl, Fukushima, and elsewhere have made local governments unwilling to tolerate the construction of a new nuclear plant anywhere nearby and have led to an increased burden of regulation. Electric utility companies have generally not been willing to make the huge investment required given the regulatory burden and political risk involved. In the last twenty years, the only new nuclear plants to come online were operated by a government agency, the Tennessee Valley Authority.[55] Two nuclear reactors are currently under construction at an existing nuclear plant in Georgia, and another two have license applications pending. The number of large solar and wind plants being planned is much larger.

Nuclear energy might become a feasible replacement for coal if there were to be an overnight change in public opinion. Even then, however, it would take considerable time for a large number of new plants to be built—probably more time than we have. Accepting for the sake of the argument that nuclear power can be made safe, it is simply not the case that it could play more than a minor role in replacing coal. Fortunately, that is not the case for energy conservation.

Conservation

If society is going to stop burning coal (and other sources of GHGs), we need to find other ways to maintain our standard of living while enabling the less developed countries to improve theirs. Most of this chapter has been devoted to possible sources of replacement energy; however, we can also achieve our goal by using less energy to produce the same goods and services. This process has traditionally been called "energy conservation." More recently, the term "energy efficiency" has been used in order to make it clear that the goal is not only to use less energy but also to maintain our quality of life while doing so. Efficiency can be measured by "energy intensity," the quantity of energy required to produce a unit of gross domestic product (GDP); the lower the energy intensity, the higher the efficiency.

Energy efficiency can be improved with little change in daily life, most simply by eliminating wasteful practices, such as leaving electric lights on and running appliances when they are not needed. It can also be improved by making appliances and engines more efficient through redesign, and most governments are promoting such changes through regulation. Other improvements involve some change in what life is like for most people without lowering the overall quality of life. These include designing buildings to use less (or no) energy, encouraging consumers to use more efficient forms of transportation, and redesigning cities so that homes, workplaces, and shopping areas are within walking distance of each other. Energy efficiency is popular because it saves money. However, when energy prices are low, there is less incentive not to be wasteful, so more regulatory action may be needed.

The International Energy Agency estimates that if efficiency improvements were pursued vigorously, energy use in 2040 could be reduced by one eighth of the amount used today. Simply implementing all government policies currently planned would lead to a reduction by 2040 of 5 percent.[56]

The full development of the various forms of renewable energy discussed above, combined with feasible improvements in energy efficiency, would be more than enough to enable the developed countries to maintain their standard of living, and the less developed countries to improve theirs, while eliminating the use of coal. However, such progress will not come automatically; it will require the adoption of suitable policies by the world's governments. In the remainder of this chapter, we will consider why such policies are needed and, in a general sense, what they should be. Later chapters will then look at attempts to achieve the needed policies through international agreements, legislation, and administrative action.

Moving Forward

Most decisions about energy use are made through the market; even public agencies, which have more freedom to pursue policy goals, are compelled by resource limitations to seek less expensive energy when they can. However, the market as it presently operates is not an efficient way to make such decisions because so many of the costs of burning coal and other fossil fuels are what economists call "externalities;" in other words, although they are true costs, neither the buyer nor the seller of energy has to pay them. The owners of a coal-fired electric generator do not have to pay for the damage to buildings, lakes, and crops caused by the sulfur in the generator's emissions nor for the damage caused by atmospheric concentration of GHGs. Government action is needed to rebalance the market. This can be done by taxing the burning of carbon or by placing regulatory limits on GHG emissions. Such limits can be made easier to implement by allowing the permitted amounts to be bought and sold in what has become known as a "cap and trade" system. In either case, the effect of policy is to make the price of coal and other fossil fuels higher so that users have an incentive to use less-polluting sources. If the extra cost of fossil fuels is sufficient, utilities and others will have a motivation to invest more in alternative energy sources. Government action may also be needed to ensure that as new sources of energy are developed, they are not simply absorbed by growing demand. The next section takes a closer look at this possibility.

The Jevons Paradox

In the mid-19th century, William Stanley Jevons noticed that improvements in coal-mining technology led to increased use of coal. The reason was simple: the improvements had reduced the price of coal.[57] Today, increasing the amount of energy produced by solar power, wind, and other alternative sources will not necessarily displace coal from the market; it may simply lower the price of energy to the point where all that can be produced is consumed. This may not happen; the price of renewable energy may fall below the point at which coal is profitable, especially if coal has to pay its full social cost. Economists have studied the possibilities without reaching a consensus.[58] However, there is at least a possibility that government action would be needed to keep fossil fuels from becoming too cheap. In the remainder of this chapter we shall look briefly at what the policy options are before moving on to see how good policies might be adopted.

Policy Options

In the larger arena of climate policy there are many noncoercive policy options, such as tax credits for renewable energy expenditures and public spending to improve mass transit. However, while such policies may increase the supply of renewable energy, they may not lead to a reduction in fossil fuels. To achieve the latter, governments have a range of choices, all involving some degree of coercion. One option is direct regulation, sometimes referred to as "command and control," where government simply sets limits on how much CO_2 can be emitted. However, this can be a cumbersome process, and governments do not always get the details right. Leaving things to the market is more efficient, but the market must be structured in such a way that GHG reductions result from market decisions. There are two ways this can be done. Most simply (and the preference of most economists) is greenhouse gas emissions could be taxed at a tax rate set to cover the value of the externalities of pollution. The resulting revenue could be used to reduce other taxes, subsidize those with low incomes so that they can continue to heat their homes and drive to work if needed, or further incentivize the development of renewable energy. Whichever choice is made, the economic result of higher prices for energy derived from fossil fuels would be the same.

A final option lay somewhere in between these two: the creation of tradable pollution rights, known as cap and trade. With cap-and-trade policies, the government makes rules to limit emissions (as with command and control), and entities that reduce their emissions below the regulated level are able to sell the surplus amount to others. Those who are not able to meet the regulatory target are able to buy credits from the first group. This market element makes the regulations more efficient, as there is an incentive for polluters to keep reducing their emissions even after they have met the regulatory target.

Conclusion

The broad policies sketched above show what needs to be done. Many more details would have to be worked out with each of them, of course.

In the final three chapters, we will look at the attempts to do so, both within the United States and in the international arena, as well as the efforts by the movement of climate activists to limit GHGs through direct action. Let us begin in the next chapter with the attempts to develop an international agreement on climate.

CHAPTER SIX

The International Politics of Coal

Climate change is a global problem, but national and local action can contribute to solving it. Without global action, however, national and local action could prove meaningless. A reduction in GHG emissions by one country could be canceled out if other countries increase their emissions by a similar amount. Moreover, until the recent appearance of cheap natural gas from hydraulic fracturing, the most carbon-rich fuel, coal, has also been one of the least expensive. As long as this continues to be the case, a country that agrees to burn less coal will pay a cost. For this reason most countries have sought a commitment, in the form of an international agreement, that if they reduce their GHG emissions other countries will do the same.

In 1968 Garrett Hardin coined the phrase "the tragedy of the commons" to argue that in the absence of a higher authority, every entity with a right to use shared resources would be motivated to use as much of them as possible, with the result that the shared resource would be destroyed.[1] Others have argued that Hardin's model does not apply to many real-world situations and that shared resources are often preserved in a sustainable fashion, even without appeal to a higher authority.[2] Nevertheless it is certainly easier for a national or local government to act to reduce GHG emissions if it has some reason to believe that others will act as well. Debate on national climate policy occurs in the context of what is happening globally. For that reason, we begin our examination of policies to reduce coal use with the attempts to arrive at an international climate agreement. These attempts date from the preparations for the

1992 United Nations Conference on Environment and Development (UNCED), popularly known as the Earth Summit, in Rio de Janeiro and continue to the present.

The Rio Earth Summit

The Rio conference was intended to mark the commitment of the nations of the world to work toward the twin goals of a safe and healthy environment and the end of poverty. Among other accomplishments, the conference established the United Nations Framework Convention on Climate Change (UNFCCC). However, the effort to reconcile the tensions between the two goals of environmental protection and economic development through the concept of "sustainability" was not completely successful; this tension was to become a dominant consideration in the subsequent debates within the UNFCCC over a climate agreement. These tensions can be better understood if see them in the context of the events leading up to the Earth Summit.

Background

The actions of the Rio summit with regard to climate had both a political and a scientific context. Politically, it was a major step in an effort that had been going on for over twenty years to unite both rich and poor nations around environmental protection. Scientifically, it came as scientists were realizing—and attempting to convey to the public—that climate change caused by GHG emissions was not a speculative future possibility but an immediate threat to the world. Let us look first at the politics.

Political History

To understand the Rio Summit, we also need to understand the first UN environmental conference twenty years earlier. The 1972 meeting took major steps, but it did not succeed in eliminating the perception that environmental problems were concerns of the rich countries, who had already polluted the world, and that the developing countries needed to do some polluting of their own if they hoped to lift their citizens out of poverty. In preparation for the 1992 summit, a commission was established to address these issues, with Gro Harlem Brundtland, a former prime minister of Norway, as chair.

The Stockholm Conference and UNEP

In 1968, Sweden proposed that there be a special United Nations conference on environmental problems. Following approval by the UN Economic and Social Council and then by the General Assembly, the UN Conference on the Human Environment was convened in Stockholm in June 1972. The conference met at a time of heightened international tension. One year earlier the General Assembly had voted, over the opposition of the United States, to recognize the Beijing-based government of mainland China, rather than the Taiwan-based Republic of China, as the holder of the Chinese seats in all UN bodies. Meanwhile, the United States was still heavily involved in the war in Vietnam, fighting against forces that had the support of many of the developing countries. These issues bubbled beneath the surface as the assembled nations strove to come up with a joint statement on environmental issues. Indira Gandhi, Prime Minister of India, summed up the basic contradiction:

> On the one hand the rich look askance at our continuing poverty—on the other they warn us against their own methods. We do not wish to impoverish the environment any further and yet we cannot for a moment forget the grim poverty of large numbers of people. Are not poverty and need the greatest polluters?[3]

The conference agreed that pollution could not be ended unless poverty was ended as well, and the new United Nations Environment Programme (UNEP) was to be based in Nairobi as a symbol of this commitment to the developing world. However, many in that world continued to see environmentalism as the rich nations' problem, and the tensions continued.[4] A follow-up conference in Nairobi held in 1987 concluded that little progress had been made.[5] Recognition of this lack of progress, and of the growing severity of environmental problems, led to the creation of the World Commission on Environment and Development, popularly known as the Brundtland Commission.

The Brundtland Report

In 1983 Javier Pérez de Cuéllar, the Secretary-General of the United Nations, asked Gro Harlem Brundtland to organize a commission to examine ways to solve environmental problems while continuing economic development. Brundtland had previously served as prime minister of Norway and had extensive knowledge of science and public health issues. The resulting body, the World Commission on Environment and Development, sought a way to define environmental preservation and economic development as a single problem, rather than as two separate and sometimes competing problems. Their solution was the concept of sustainable development, defined as "development that meets the needs of the present without compromising the ability of future generations to meet their own needs."[6] This language answered the charge that environmental protection was a purely aesthetic and recreational concern of rich people who wanted to enjoy their nature preserves and yachting trips; instead, it was a matter of preserving society's ability to provide a decent standard of living for everyone on a long-term basis. This was to be a central theme of the Rio conference, although considerable tension still existed when it came to the specifics of how to make sustainable development work. Meanwhile, a growing realization among scientists that climate change was a serious problem was beginning to enter public consciousness.

Scientific Background

The development of the scientific understanding of climate change is discussed in detail in chapter 3. James Hansen delivered his dramatic testimony to the U.S. Senate in 1998. The IPCC issued its First Assessment Report in 1990, concluding with certainty that the earth was being warmed by the greenhouse effect, with CO_2 contributing about half of the increased warming, and with less certainty that with no change from business as usual the average global temperature was likely to increase by 0.3°C every ten years. As a result, climate change took its place among the environmental issues requiring global action but was not yet seen as extremely urgent. It thus became one of several issues to be acted upon by the conference.

Creation of the UNFCCC

Major international conferences have several purposes. They demonstrate commitment to a common purpose, they give individual nations and their leaders a chance to show their own commitment as well as highlight any concerns they have, and they can generate enthusiasm for new international agreements and other initiatives. The Rio conference served all of these purposes. National leaders from 171 countries took part, with a total of 178 countries represented. There was a great deal of discussion about the comparative responsibility of rich and poor nations for action to save the environment, but this debate did not prevent the issuing of several important documents. Among these was the UNFCCC.[7]

How the UNFCCC Came To Be

The preparation for a major international conference begins long before the actual event. Since the length of the actual conference is too short to negotiate very much, most of the details of any agreement must be worked out in advance through a series of preparatory meetings among the parties. The actual conference serves to provide both a deadline by which an agreement must be reached and a highly visible platform from which the agreement can be launched. While setting such a firm deadline probably does make it easier for an agreement to be worked out, it also gives power to any negotiating parties who have strong objections but whose participation is seen by others to be essential.

In this case a General Assembly resolution on December 21, 1990, had created the International Negotiating Committee for a Framework Convention on Climate Change, which then met first in February 1991.[8] It quickly became clear that there was a major disagreement between most European nations, which wanted to combine a framework agreement with a specific goal for GHG reduction (most likely a return to 1990 levels by 2000), and the United States (and to some extent Russia), which wanted to omit specific goals from the agreement, leaving them to be negotiated at subsequent meetings.

In February 1992, the negotiations on this issue were at an impasse. Several European nations sent delegations to Washington to try to persuade the United States to accept targets, but they did not succeed. Then in April, President George H. W. Bush talked with Helmut Kohl, Chancellor of Germany, and made it clear that he would attend the conference only if any specific commitments were left out of the agreement. Kohl agreed, the final draft of the treaty was altered accordingly, and Bush announced his plans to go to Rio.[9]

The result was a treaty that was almost purely formal. It declared the intention of the signing parties to return GHG concentration to "earlier levels" by 2000, a phrase that some interpreted as meaning the levels as of 1990 but others were free to ignore. There were no commitments at all for subsequent years. Essentially the treaty left all disagreements unresolved but created a mechanism to attempt to work them out in the future.[10]

How the UNFCCC Works

The UNFCCC is a framework convention, which is a fairly common structure in international law. Essentially, it is an agreement to establish a framework through which future, more specific agreements can be negotiated. In this case, the signing nations agreed that it would be desirable for the atmospheric concentration of GHGs to be reduced to the level that pertained in 1990. The conflict over the roles of rich and developing countries was addressed by two provisions. First, the signing countries were classified into "Annexes." Those listed in Annex I were considered to be developed, while those not in Annex I were considered to be developing. Those Annex I countries that belonged to the Organization for Economic Cooperation and Development (OECD) were also listed in Annex II. Applying the principle (also developed at UNCED) of "common but differentiated responsibility," Annex I countries were to be required to reduce their GHG emissions first; non-Annex I countries would agree to act only after the Annex I countries had made reductions. Second, a mechanism was to be created for Annex II countries to contribute money and technology to help non-Annex I countries reduce their emissions in the future.[11]

In addition to stating the above-mentioned goals, the UNFCCC also prescribed a process for reaching further agreement. The member countries ("states parties" as they are called) were to meet in a Conference of the Parties (COP) once every year for the purpose of further negotiations, and a UNFCCC Secretariat was created to provide support for the annual COPs and other conferences, to facilitate communication among the parties, and to give technical support. (As of 2018 the Secretariat employs about 450 people based at its headquarters in Bonn.[12]) President Bush signed the treaty at the end of the Rio Conference on June 12, 1992, and he U.S. Senate ratified it on October 15 of that year. The treaty was to enter into force once fifty countries had ratified it, which took place on March 21, 1994. The first COP met the following March in Berlin, with subsequent meetings every year since then. The first major agreement was negotiated at COP 3 in Kyoto in 1997 and became known as the Kyoto Protocol. We shall now take a more detailed look at this protocol.

Development of the UNFCCC

The climate agreement signed at Rio was a step forward, but it needed to have specific content added before it would be meaningful. This was to be worked out by the annual Conferences of the Parties (COPs), meetings of all the signatory nations. Several major issues had to be resolved before progress could be made. First, in disagreement with most European countries, the United States did not want to agree to specific, obligatory amounts of GHG emissions reductions. By the terms of the U.S. Constitution, any treaty had to be ratified by a two-thirds majority of the Senate before it would constitute a national commitment, so the Clinton administration was reluctant to agree to anything that could not get the approval of enough senators.

A second issue involved the concept of "common but differentiated responsibilities." This phrase was included in the Rio Agreement to indicate that every country should do something to help prevent country change, but that the developed countries, which had become rich by burning lots of fossil fuels, should do more than the developing countries. The phrase was quantitatively ambiguous by intention, so that everyone could agree to it, but any meaningful agreement on climate would have to specify the details of each national obligation.

A third issue had to do with forests. The Rio conference had reached a forest agreement as well as a climate agreement, but doing so had required some major differences to be papered over. The developing countries were strongly resistant to the idea that their forests should function as carbon sinks for the world. They feared that such a decision would undermine their national sovereignty, and more concretely that it would deprive them of a valuable national resource, which could otherwise be sold to fund development. The developed countries, on the other hand, had large groups of environmentally-conscious citizens who wanted to save virgin forests for their own sake, as well as fossil fuel businesses who hoped to use the preservation of forests in other countries as an offset to their own continued GHG emissions. All of these issues had to be resolved in the ongoing negotiations for a real climate treaty.

First Attempts

The first two COPs, in Berlin (March 28–April 7, 1995) and Geneva (July 8–19, 1996), were mostly concerned with establishing the framework for future meetings and ensuring that adequate technical support would be available. However, at the Geneva meeting, the head of the U.S. Delegation, Undersecretary of State for Global Affairs Tim Wirth, made a strong statement accepting the findings of the Second Assessment Report of the IPCC and agreeing that the United States and other countries needed to make specific commitments to GHG reductions.

Wirth, a former Democratic senator from Colorado, had a long record of climate activism. As a senator, he had organized the 1988 hearings where James Hansen had announced his conclusion that human activity was changing the climate; Wirth later acknowledged that, in order to heighten the effect of the hearings, he had chosen to schedule the hearings on a day that was likely to be very hot and had opened all the windows in the hearing room the night before so that the air-conditioning was not operating. Clinton had given him the lead responsibility in the climate negotiations, a position he used to an advantage. Wirth's statement in Geneva helped set the stage for the pursuit of a major agreement at the next COP, to be held in Kyoto in 1997.[13]

Kyoto and Its Aftermath

The third Conference of the Parties of the UNFCCC (COP 3) was scheduled for December 1–10 in Kyoto, Japan. The members had agreed that they should try to reach an agreement to reduce GHG emissions at this meeting, and preparatory meetings had clarified a number of issues that would have to be resolved. In addition to the tension between rich and poor nations, discussed below, there were some disagreements among the rich nations. These disagreements included both the amount by which emissions should be reduced and what would or not count as a reduction. On the former point, the European Union (EU) had proposed a reduction by 2010 of 15 percent below the emissions levels that had prevailed in 1990, while the United States wanted a reduction only to 1990 levels to be attained somewhere between 2008 and 2012. In either case, every developed country (referred to in the UNFCCC as Annex I) would be obligated to reduce its emissions by the same percentage. As for what would be considered a reduction, there were several issues. The EU wanted to be treated as a single entity, so that smaller reductions by some EU members could be made up for by larger reduction by others; other member states objected to this. The United States in turn wanted to be able to count the preservation of forests as carbon sinks as reductions.

The United States also wanted a system of global carbon trading, whereby a nation could meet its target either by reducing its own emissions or by buying the reductions achieved by another country.[14] Both the United States and the European Community accepted the basic concept of carbon trading, but there were major disagreements about the rules by which it should be implemented. In particular, the European countries wanted to exclude trading with Russia. Since the Russian economy had collapsed following the end of Russia's communist system, Russia's GHG emissions had already fallen below 1990 levels. Thus, Russia potentially had surplus reductions that it could sell, a phenomenon that became known as "hot air." If the United States or another country met its quota by buying this hot air, no actual reductions would have been achieved.[15]

The developing countries (non-Annex I) objected both to carbon trading and to giving credit for forest preservation. They were unwilling to give up their sovereign right to develop their own forests, and they feared that carbon trading would enable rich countries to buy emissions reductions from the developing world, with possible accompanying economic losses. More generally, the developing countries insisted that they would not commit to any reductions until the rich countries had reduced their own emissions.[16]

A further difficulty arose from the Constitutional rules in the United States about treaties, which must be ratified by a two-thirds vote of the U.S. Senate in order to take effect. This rule made it difficult for the U.S. negotiators to seek a compromise, since compromise might mean that the treaty could not be ratified. In particular, the idea that China and India, major emitters of GHGs, were non-Annex I countries potentially not included in the initial reduction targets was very unpopular in the Senate. All these issues and more were on the table as the conference began.

Adoption of Protocol

COP 3 convened in Kyoto on December 1, 1997, with several issues still on the table. The industrialized countries disagreed about how much each of them would have to reduce its GHG emissions by and about how to measure the reductions. The United States sought a less stringent level of reductions for itself, equal to but not lower than 1990 levels, and it wanted (together with Canada and Australia) to count the impact of its forests toward their reduction targets, while most of the rest of the world thought that this was just an excuse for doing less. The United States also sought a nonquantitative commitment from developing countries that they would reduce their GHG emissions in the future, but this provision was dropped after China and India made it clear that they would not sign the agreement if such a provision were included.[17]

In addition, the United States and the bloc of developing countries each had an issue that it was not willing to compromise. For the United States, this was the adoption of an emissions-trading scheme that would let it meet part of its commitment by paying for GHG reduction projects in other countries. The developing bloc, on the other hand, was not willing to agree to anything until it saw the industrial countries actually making reductions.[18]

These issues remained unresolved until the last three days of the conference when the delegates were joined by ministerial-level officials,
including Vice President Al Gore of the United States. Gore agreed to
accept a more stringent emissions-reduction target for the United States,
to 7 percent below 1990 on average during the years 2008 to 2012. After
an extra day had been added to the conference, it was agreed to set up
what became known as the Clean Development Mechanism, to facilitate
the transfer of new technology to developing countries and in principle to
allow for emissions trading, but to leave the details of how that would
work to a future meeting in November 1998. U.S. negotiators stated that
the United States would not sign the treaty until emissions trading had
been approved. Ominously, Senator Chuck Hagel (R-NE) predicted that
the Senate would not approve the treaty.[19]

U.S. Withdrawal

Hagel's prediction was borne out. President Clinton signed the agreement on November 12, 1998.[20] However, he never sent the agreement to
the Senate for ratification, knowing that it was certain to fail if he did.
Two years later, Vice President Gore lost the presidential election to the
Republican candidate, George W. Bush; a few weeks after his inauguration, Bush formally withdrew the United States from membership in the
Kyoto Protocol, stating that he opposed it "because it exempts 80 percent
of the world, including major population centers such as China and India,
from compliance, and would cause serious harm to the U.S. economy."
Bush also argued that scientific knowledge of "the causes of, and solutions
to, global warming"[21] was insufficient.

The U.S. withdrawal was a major setback to the process of negotiating
a climate agreement; however, it did not destroy the Kyoto Protocol. Several other industrialized nations expressed their intention to continue to
follow the treaty's provisions, and efforts were made to bring it into force
without U.S. participation. The treaty had provided that it must be ratified by fifty-five countries and that this number must include Annex I
countries representing at least 55 percent of total GHG emissions from
Annex I countries as of 1990; non-Annex I countries could sign, but their
emissions did not count toward the 55 percent threshold because they
were not obligated to make any reductions during the protocol's first
commitment period (1998–2012). Since the United States was responsible for 21.6 percent of global emissions, it was difficult, but not impossible, to reach that threshold without U.S. participation.[22] By 2004 the
figure was 44 percent, and the 55 percent goal was attainable if Russia

would join. As mentioned earlier, it was fairly painless for Russia to do so as they had already reached their reduction target due to economic collapse; however, they drove a hard bargain, finally agreeing to ratify the treaty if the EU in turn would support their application for membership in the World Trade Organization (WTO). The agreement was made, Russia ratified the Kyoto Protocol on November 18, 2004, and the protocol went into force on February 16, 2005.[23]

The Kyoto Protocol has its own Conference of the Parties, an annual meeting of representatives from all the countries that have approved the protocol. Since some countries, notably the United States, had agreed to the UNFCCC but not the Kyoto Protocol, the situation could have become even more cumbersome than it already was. To avoid this, it was agreed from the beginning that the annual COP of the UNFCCC would also serve as the COP of the Kyoto Protocol. All UNFCCC member states would be present for the annual meeting, but those states that had not adopted the protocol would serve as observers when specific items about the protocol were taken up. Among other results, this allowed the United States to stay involved in discussion of future steps to be taken, including renewal or replacement of the Kyoto Protocol at the end of its initial period in 2012.[24] COP 11 of the UNFCCC/COP, serving as the meeting of the Parties to the Kyoto Protocol (CMP 1), met in Montreal November 28–December 9, 2005, with further meetings in Nairobi in 2006, Bali in 2007, Poznań in 2008, and Copenhagen (COP 15/CMP 5) in 2009.

The agenda of each COP included reports from each member state on progress toward its commitments, working out the details of such matters as emissions trading and the Clean Development Mechanism and, perhaps most importantly, attempting to reach agreement on how to move forward after 2012. Progress was difficult. The United States insisted that China and India would have to agree to make specific GHG emissions reductions in any future agreement; China, in turn, was unwilling to make any commitments unless the United States did so first. U.S. president George W. Bush had expressed doubt about the danger of climate change, but even under his predecessor, the more environmentalist Bill Clinton, the United States had been unable to ratify the Kyoto Protocol. Moreover, as will be discussed further in the next chapter, attempts to pass a climate law in the U.S. Congress continued to fail. Developing nations were very reluctant to make any commitments unless the United States first showed that it was willing to do so.

The Bali meeting in 2007 did agree to a process for reaching a new agreement. The COP created two working groups, one under the UNFCCC and one under the Kyoto Protocol, and set COP 15/CMP 5 in Copenhagen as the target for reaching an agreement. The two working groups met eight times and produced two hundred pages of draft text, but with disagreements over many important points, including whether to have one agreement or two (reflecting the UNFCCC and the Kyoto Protocol) and how to bring the developing countries into the GHG reduction process. China continued to insist that it would not accept any binding commitments to reduce its emissions.[25]

Climate issues did not play a big role in the 2008 presidential campaign; attention was focused on the Great Recession, which had begun in 2007, as well as the ongoing wars in Iraq and Afghanistan and Obama's proposal for a national health care plan. Nevertheless the winning candidate, Barack Obama, clearly accepted the prevailing scientific understanding of the problem of climate change.[26] Moreover, he was a strong believer in the value of international cooperation. For both those reasons, environmentalists hoped that he would bring the United States back into the UNFCCC process in a positive role as well as seek to build GHG reductions into U.S. domestic policy. There were high hopes, therefore, that the U.S. role in the climate negotiations would become more positive and significant steps could be taken in the Copenhagen meeting.

Copenhagen

A few days after his electoral victory, President-Elect Obama sent a four-minute video message to a climate summit of state governors hosted by Governor Arnold Schwarzenegger (R-CA). In the message Obama condemned the inaction of the Bush administration, promising that it would change and that he would open "a new chapter in America's leadership on climate change." He pledged to pursue a federal cap-and-trade system and to provide $15 billion per year to support alternative energy sources. He closed with a message of support for the next COP, which was to meet in Poznań, Poland, December 1–12, 2008. Obama explained that he would not yet be president at the time of the meeting, but he pledged his support to move ahead with an agreement once he took office.[27] The Poznań meeting (COP 14/CMP 4) decided to have the proposed draft of a new climate agreement available in time for a UNFCCC meeting in Bonn in June 2009, with the goal of adopting it in Copenhagen in December. Given Obama's public comments, the Copenhagen conference was anticipated by many with great optimism.

High Hopes

President Obama took office with an ambitious legislative agenda. Major goals other than a climate bill included an economic stimulus to help end the recession and the creation of a national health care system. Both were highly controversial, as was the climate bill, but the Obama administration and its Congressional allies showed a great deal of strength. The American Recovery and Reinvestment Act (ARRA), an $800 billion stimulus package combining tax cuts and spending increases, was signed into law on February 17, 2009, less than a month after Obama's inauguration. Among the law's spending provisions were grants to support renewable energy, high-speed railroad lines, and other climate-friendly projects. The House of Representatives passed a cap-and-trade bill, the American Clean Energy and Security Act of 2009, on June 26, by a vote of 219–212. As it turned out, the bill was killed in the Senate; at the time, however, it was seen as generating further momentum for success in Copenhagen. Jennifer Haverkamp of the Environmental Defense Fund told the *Guardian*, "I think it will have a very positive impact on the Copenhagen process because the international negotiations have largely been stymied by countries waiting to see what the U.S. will do."[28]

Most environmentalists hoped that Obama's support for climate action would lead the Copenhagen conference to adopt a new, stronger version of the Kyoto Protocol that increased the rate of GHG reductions and extended the obligation to make reductions to China, India, and other developing countries. However, there were reasons for caution. Most important, while there was hope that the president's allies in the Senate could overcome a filibuster (requiring sixty votes) to pass a cap-and-trade bill, there was very little possibility that they could secure the sixty-seven votes needed to ratify a new treaty. In addition, Obama continued to support the long-held U.S. view that the Kyoto structure of imposing top-down requirements for reductions should be replaced by a bottom-up "pledge-and-review" structure whereby individual countries promised to make specific reductions, which would then be reviewed at periodic conferences.[29] With these goals in mind, Obama hoped to secure an agreement that would not constitute a treaty so that it would not require Senate ratification. This played out in unexpected and controversial ways in Copenhagen.

Apparent Failure

COP 13 had agreed to the Bali Action Plan, according to which two Ad Hoc Working Groups (AWGs) would develop the draft of a new agreement in time for it to be adopted by COP 15 in Copenhagen. As the Copenhagen meeting opened, the draft existed but with several issues still in disagreement. These included whether to extend the Kyoto Protocol as a binding agreement or to replace it with a different structure, whether the target should be to limit the increase in global mean temperature to 2° or 1.5° Celsius, what specific reductions Annex I countries should be held to, whether non-Annex I countries should also be required to achieve reductions, whether any emissions reductions by non-Annex I countries should be independently verifiable, how much money should be provided for developing countries to pursue adaptation measures, and where such money would come from.[30]

The issues seemed intractable. On the one hand, China, which had now become the largest emitter of GHGs, refused to accept any specific binding GHG reduction target, a position in which it was supported by India and Brazil, or to accept international monitoring and verification of its reductions. On the other hand, the United States was offering a GHG reduction of 17 percent below 2005 levels by 2020, the amount specified by the Waxman-Markey bill. Since other countries' commitments were based on the lower 1990 levels, and since it had become clear that Waxman-Markey could not pass the Senate, the U.S. pledge was not seen as credible.[31]

Faced with these difficulties, President Obama did not come to the conference with the other 120 heads of government but rather arrived on the last day. After delivering a speech chiding the attendees for not having reached an agreement, Obama spent the next thirteen hours in small meetings with other national leaders. When Obama went to the room assigned for a second face-to-face meeting with Chinese Premier Wen Jiabao, he found Wen already meeting with the leaders of Brazil, India, and South Africa. The other three leaders stayed on for the Obama-Wen meeting, and the five proceeded to work out the twelve-paragraph document that became known as the Copenhagen Accord.[32] Obama then called a press conference, announced that an agreement had been reached, and boarded Air Force One to fly back to Washington.

The immediate reaction of other delegates to the announcement of the agreement was largely hostile, in part because Obama had announced the agreement before it had been approved and in part because it had not been produced by the AWGs (in which all member states participated) but instead by a small group of states that presented it on a take-it-or-leave-it basis. Moreover, it was nonbinding and left the amount of GHG reduction by each country up to the voluntary action of that country. Several countries remained adamantly opposed to the agreement, including Sudan, Tuvalu, Venezuela, Nicaragua, Cuba, and Bolivia. Since the UNFCCC requires consensus, the conference voted only to "take note" of the Copenhagen Accord.[33]

Many environmentalists were shocked by this development. Bill McKibben, for example, called the accord a "non-face-saving pact" among a "cartel of serious coal-burners [that] laid out the most minimal of frame-works."[34] Lumumba Stanislaus Di-Aping, the Sudanese delegate and chair of the bloc of less developed nations (G-77), declared, "What has happened today confirms what we have been suspicious of that a deal will be imposed by United States, with the help of the Danish government, on all nations of the world."[35] Eventually 120 countries did sign on to the accord, but many feared that it marked the end of the UNFCCC. As McKibben put it, "If you want to despair, that's certainly a plausible option."[36] Nevertheless, the accord did contain some possible seeds for further progress. These will be examined in the next section.

How the Copenhagen Accord Worked

The new accord took a very different approach from that of the Kyoto Protocol. Rather than putting national GHG reduction goals into the agreement, the accord left it to each country to declare what reductions it planned to achieve. The accord did specify that the goal was to hold the increase in mean global temperature to 2° Celsius or less. Both Annex I and the wealthier non-Annex I countries were expected to submit targets but with an understanding that for the latter these were subservient to the needs of development and poverty eradication. Participation was voluntary for less developed countries and small island developing countries. Once a country had set targets, its annual performance in achieving those targets was to be measured, reported, and verified on a biennial basis. The developed countries were to provide financial support for adaptation and mitigation efforts in developing countries, at a level to reach about $30 billion by 2010 to 2012, with higher levels in later years. The effectiveness of the accord was to be reassessed in 2015.[37]

Two details of the accord had been especially controversial. First, the goal of limiting global temperature increase to 2° Celsius would not save some low-lying island states from complete inundation. Tuvalu, supported by other island states, had made this point insistently during the COP but had not carried the point. The only concession was that the 2015 reassessment would include "consideration of strengthening" the goal to 1.5° C. Second, the United States and Obama personally were on record asserting that transparency in the form of international measurement, reporting, and verification of national goals was essential to the success of the accord. China and India, on the other hand, insisted that international monitoring would violate their sovereignty, and was unacceptable to them. A compromise was worked out during the Friday evening face-to-face meeting: mitigation supported by international funding would be subject to international measurement, reporting, and verification, but for efforts a country funded by itself—and this was true of most projects in China and India—this phrase was replace by "international consultations and analysis" with an understanding that national sovereignty would be protected. This wording enabled both the United States and China to sign the accord.[38]

The more general criticism, as represented by the quotations from Bill McKibben and Lumumba Di-Aping earlier, was that the accord did nothing. No nation was required to meet any particular reduction target nor was there any enforcement for the targets they set for themselves. Two months later, the Potsdam Institute for Climate Impact Research reported that the pledges made so far left the world "heading for a global warming of over 3°C above preindustrial levels by 2100."[39] I was pessimistic myself; in a chapter published in 2014, I asserted, "Although the UNFCCC has limped on, with subsequent meetings in Cancún, Durban, and Doha, it no longer seems capable of bringing about the reductions that are needed."[40]

Such pessimism proved premature. The 2015 meeting of COP 21 in Paris produced an agreement that most considered a major step forward. The form of that agreement grew directly out of the Copenhagen Accord.

Paris

As had been the case with Copenhagen six years earlier, the Paris COP was approached with high hopes by environmentalists. This time, however, those hopes had a firmer foundation. Most importantly, the United States had finally shown that it had the will and the capacity to reduce its own GHG emissions.[41] China, too, had begun to slow the growth in its emissions, in part in order to deal with a domestic crisis in the old type of air pollution. Almost all world leaders agreed that there was a need to act, and many subnational governments were taking action on their own, without waiting for an international agreement. This time, the expectations were not disappointed—at least not immediately.

Initial Steps

The UNFCCC is a large and cumbersome organization. It has 197 parties and is divided into several blocs with different interests. Many of these blocs, and several of the individual members, are effectively able to veto any agreement in that their refusal to participate would make the agreement worthless. Understandably, progress is slow and can often seem not to occur at all. Although the annual COPs have a lot of routine work to do, such as receiving and summarizing data from the member states, substantive progress can be limited to little more than setting the agenda for the next meeting.

When one steps back a bit, however, it can be seen that change is occurring, if slowly. The annual COPs, together with the more frequent meetings of various AWGs between the Copenhagen meeting in 2009 and the Paris meeting in 2015 managed to reach consensus on several issues and to frame others in a way that allowed them to be decided in Paris. The Cancún conference in 2010 made the Copenhagen Accord, which had been "taken note of" but not accepted in Copenhagen, a formal part of the UNFCCC process, but it did not extend the pledges member countries had made past 2020. It also did not decide whether to extend the Kyoto Protocol, with its top-down imposition of GHG reduction targets, beyond 2012. The next COP, in Durban in 2011, adopted the Durban Platform for Enhanced Action. The essential elements of this agreement were that the Annex I countries agreed to extend their Kyoto commitments through 2020, while the large, fast-growing non-Annex countries (Brazil, China, India, and South Africa) agreed to aim for a new agreement with legal force to apply "to all parties" (but not necessarily uniformly) from 2020 on. In addition, the small-island states and other low-lying countries won agreement to put the issue of holding average global temperature increase below 1.5°C on the table. The Ad Hoc Working Group on the Durban Platform (ADP) was created to work on the details in more frequent meetings.[42]

Subsequent COP meetings in Warsaw (2013) and Lima (2014) agreed that states should submit the Intended Nationally Determined Contributions (INDCs) called for by the Copenhagen Accord in advance of the Paris meeting (180 did so), and they established standards for how the INDCs should be described. After fifteen meetings, the ADP completed its work with the release of the draft of a new agreement in February 2015. Many issues remained undecided, but the basic form of what was to become the Paris Agreement was now clear.[43]

Agreement in Paris was made easier by some developments outside the UNFCCC process. Perhaps most important, the United States had finally adopted a strong program to decrease GHG emissions. Faced with blockage in Congress, Obama had adopted a regulatory approach, helping the EPA issue strong new rules on emissions from motor vehicles and electric power plants; we shall look at this in more detail in chapter 7, but note here that this action restored the ability of the United States to assert its leadership in global climate policy.[44]

Second, Obama and Chinese President Xi Jinping met in November 2014, and again in September 2015, expressing their mutual support for a strong Paris agreement. In the 2014 meeting, Obama announced the U.S. INDC: a 26–28 percent reduction from 2005 levels by 2025, and Xi committed China to halting the growth of its CO_2 emissions by 2030. The 2015 joint statement reiterated their mutual support for and intent to provide leadership to the Paris COP. This pledge was important because both states had behaved somewhat roguishly in Copenhagen; Obama had not attended until the last day, while Premier Wen Jiabao of China sent lower ranking officials to several of the meetings instead of attending himself. There were many other issues dividing the United States from China, so their cooperation on climate was a very positive step.[45]

Finally, persuasion of national leaders may have been a factor in the success of the Paris COP. Radoslav S. Dimitrov, who was a delegate to the talks for the European Union, argues that the continued presentation since 2005 of arguments that nations could experience green growth with a "win-win" approach to GHG control had "changed perceptions of the economic benefits of climate policy" leading to a "spate of domestic developments around world."[46] To a considerable extent, national leaders and diplomats had come so see GHG reduction as a benefit more than a cost.

A final factor was the personal skill and commitment of the chair of the conference, French Foreign Minister Laurent Fabius. Fabius held a very large number of individual and small group meetings with representatives of parties who had major disagreements with each other and/or with the proposed text of the agreement, reached compromises that were kept secret until the end, and then presented the conference with a final draft on a "take-it-or-leave-it" basis. There was one final moment of tension when the United States objected that the word "should" had been changed to "shall" in a statement about national emissions reductions, but after ninety minutes of tense private meetings, this change was announced to have been a typographical error, and the agreement was adopted. Several participants and observers have reported that the conference would not have succeeded with Fabius's leadership.[47]

Issues

Despite these favorable preconditions, several major issues remained on the table when the delegates arrived in Paris. First, how binding would the agreement be? The Kyoto Protocol had been a treaty, requiring the parties to achieve the emissions reductions it specified. The Copenhagen

Accord had not been a treaty and did not require signatories to do any-thing other than by their own choice. Most countries wanted the new agreement to be a binding treaty, but this choice faced major obstacles. First, under U.S. law any agreement regarded as a "treaty" would require ratification by the U.S. Senate, and this was politically impossible. How-ever, the definition of "treaty" in international law is looser than that in U.S. law; an agreement that did not require new spending or impose new binding emissions reductions could be treated in the United States as an executive agreement, not subject to ratification. In addition, not only the United States but also India and China were not willing to have specific emissions reductions imposed on them by the force of law.

The answer arrived at in Paris was to make some actions compulsory and others optional. Countries would be required to submit their GHG reduction targets as "Nationally Determined Contributions," to do so in a form specified by the agreement in order to meet the requirement of transparency, to report regularly on progress toward meeting those tar-gets, and to revise their targets every five years. With regard to the last requirement, there was a strong suggestion that the revision should be toward greater reductions, but there was room for flexibility.[48]

A second issue was how ambitious the agreement should be. The goal could have been expressed either as the target atmospheric concentration of GHGs or as a target limit in temperature rise. Based on increasingly alarming projections by the IPCC, the parties agreed to aim for a target of global temperature rise "well below 2°C." In addition, the agreement specified that the goal was to achieve "global peaking" as soon as possible and to make reductions "so as to achieve a balance between anthropo-genic emissions by sources and removals by sinks of GHGs in the second half of the century." The omission of specific reduction levels or target dates made it possible for all parties to accept this language.[49]

The final remaining issue was how to apply the principle of "common but differentiated responsibilities" in the UNFCCC (commonly short-ened to "differentiation"). Several countries that were "non-Annex I" in the Kyoto Protocol had experienced rapid growth, not only in their economies but in GHG emissions. China was now the world's largest emitter of GHGs by volume (though not per capita); South Korea and Mexico had joined the OECD; and Singapore and Qatar were now among the richest countries in the world. The United States, in particu-lar, was not willing to accept an agreement that did not impose emis-sions limits on China; moreover, it would be difficult to achieve the target for limiting global warming if these countries did not make a contribution. On the other hand, the non-Annex I countries saw differentiation as a moral issue: since the Annex I countries had caused the problem of climate change, they should bear the brunt of the responsibility for controlling it. Moreover, most of these countries still had high levels of individual poverty, and wanted to devote their resources to dealing with that.

This issue was resolved by eliminating mention of the annexes and imposing obligations on all parties. However, these obligations were formal: each party had to submit its Nationally Determined Contributions and to report them in ways specified by the agreement, but the specifics of its NDC were left for each party to determine. Basically, while the agreement imposed general obligations, each country was free to differentiate itself.[50]

With these issues resolved, Fabius's draft was accepted by COP 21 on December 12, 2015. It was opened for formal signatures on April 22, 2016. The agreement provided that it would enter into force thirty days after fifty-five countries representing 55 percent of world GHG emissions had ratified it; the threshold was reached on November 2016. As of July 2018, it had been ratified by 179 countries.[51]

The Paris Conference provided a platform for other actions and alliances to combat climate change. Prominent examples include the Compact of Mayors, which pledged 450 cities (as of 2016) to reduce their GHG emissions; the Breakthrough Energy Coalition, an alliance of investors, including Bill Gates, Jeff Bezos, and Mark Zuckerberg, to support green energy; and the Paris Pledge for Action, an opportunity for nonstate actors to pledge to contribute to GHG reduction and signed by thirteen hundred organizations before signatures were closed.[52]

The United States signed the Paris Agreement on the first day and submitted a formal ratification on September 3, 2016. However, Donald Trump campaigned against the agreement and pledged to withdraw the United States from it if he was elected. On June 1, 2017, Trump announced that the United States would withdraw on the first permissible date, four years after the agreement went into force. That will be November 4, 2020, one day after the next presidential election.[53]

The Aftermath

Despite the pending withdrawal of the United States, the Paris Agreement remains in force. Meetings of the Conference of the Parties serving as the Meeting of the Parties to the Paris Agreement (CMA) took place in Marrakech (November 15–19, 2015) and Bonn (November 6–17, 2017) with a third session planned for Katowice in December 2018. Other world leaders have united in rejecting President Trump's call to renegotiate the agreement and in pledging to move forward with its work.[54]

The effect of the Trump administration on the international climate regime, as well as other climate policies, will be examined more closely in chapter 8. The next chapter will look more closely at the developments in U.S. climate policy that helped lead to the breakthrough in Paris.

CHAPTER SEVEN

Coal and Climate in the Obama Administration

The 21st Conference of the Parties to the UNFCCC (COP 21) would probably not have arrived at the Paris Agreement had the United States not adopted a strong plan to reduce its own GHG emissions. Doing so was one of the major achievements of the Obama presidency, but it was not easily done. Moreover, that plan is now under threat from the Trump administration. In order to understand this, we need to know something about the development of climate policy over the past few decades.

Before Obama

Although the United States signed the Kyoto Protocol during the Clinton administration, President Clinton never submitted it to the Senate for ratification since he knew that it would fail. Following the election of George W. Bush, who proceeded to withdraw the United States from the Kyoto Protocol, there were attempts to control GHGs through legislation, through regulatory initiatives, and through the courts. There were also significant climate initiatives at the state and local level. For the most part, with one significant exception, efforts to create federal policy failed.

Congressional Action

There were 110 proposals for climate action during the 109th Congress (2005–2006), followed by 235 in the 110th.[1] Many of these were amendments to other bills or proposed nonbinding resolutions. However, a significant bill, the Climate Stewardship Act, was introduced in 2003 by Senators John McCain (R-AZ) and Joseph Lieberman (D-CT).[2] Similar bills were introduced in 2005 and 2007. The 2003 version would have set up a cap-and-trade system for GHG emissions from the electric, transportation, and manufacturing sectors of the economy, with the cap for 2010 set at the emissions level of 2000. Residential and agricultural emissions would have been exempted, but the included sectors produced 85 percent of U.S. GHG emissions. The bill was defeated in the Senate, 43–55, on October 30, 2003. Many of those who voted against it, a group that included several Democrats, cited concerns that the proposed law would hurt the economy and threaten jobs.[3] The bill was renamed the Climate Stewardship and Innovation Act and reintroduced in the next Congress. McCain sought to attach the bill as an amendment to HR 6, the Energy Policy Act of 2005, but it was voted down 38–60 on June 22, 2005. A third version of the bill was introduced in 2007 but died in committee.

Although the support for a climate bill in the Senate was drawn from both parties, those supporters did not constitute a majority. The House of Representatives was even less promising; the House is more tightly controlled by the majority party, and the Republican majority that lasted from 1995 to 2006 did not support action on climate. The Democrats regained the House majority in the 2006 election, but they were not united on climate. A number of members from coal-producing states, such as Rick Boucher of Virginia and Nick Rahall of West Virginia, held important subcommittee positions and used them to emphasize research into carbon capture and storage rather than reduction of fossil fuel use; as we saw in chapter 4, this was the position advocated by the coal industry.[4] After months of debate, a climate bill was introduced in October 2008 by Boucher and John Dingell (D-MI), the chair of the House Energy and Commerce Committee. Dingell was known for his defense of the automobile industry's interest in limiting emissions controls on cars, and Boucher, as stated already, was close to the coal industry. The resulting bill proposed very moderate reductions over a long period, to 80 percent of 2005 emissions by 2050, by requiring new coal plants to use carbon capture and storage.[5] Whatever the merits of the bill, there was not time to consider it prior to the fall election and the end of the 110th Congress. Meanwhile, the Bush administration was attempting to move climate policy in a different direction.

Executive and Regulatory Action

President George W. Bush cited two reasons for his decision to withdraw the United States from membership in the Kyoto Protocol. He asserted that the scientific understanding of climate change was not certain enough to warrant taking costly action and that the required cuts in GHG emissions would indeed be costly, driving up the price of electricity and threatening to undermine the American way of life.[6] However, regardless of the doubt he expressed about the science, he continued to develop his own conservative version of climate policy. Many environmentalists considered Bush's climate proposals to be inadequate to the need; however, Bush preferred to argue that his strategy was a better solution to the problem rather than to deny that the problem existed.

Bush's first initiative in this area was the National Energy Policy. Bush had appointed a task force, the National Energy Policy Development Group, chaired by Vice President Dick Cheney, shortly after taking office. The task force reported its proposed National Energy Policy on May 17, 2011. The main focus of the policy was not on climate but on increasing energy supplies through accelerated drilling for oil and gas, including drilling in the Arctic National Wildlife Refuge. However, it did include subsidies for the development of CCS technology as well as a suggestion that GHG emissions growth could be slowed if new gas-powered electrical plants replaced coal-powered plants. The proposals to increase oil drilling drew most of the public attention.[7]

President Bush's National Energy Policy was followed in February 2002 by his announcement of a climate strategy. The key element in this strategy was an emphasis on cutting "greenhouse gas intensity," defined as the amount of GHG emissions required to produce a given amount of gross domestic product (GDP). Bush's goal was to reduce GHG intensity from 183 million metric tons per million dollars of GDP in 2002 to 151 million metric tons per million dollars in 2012. This was primarily a call for voluntary action by industry (not unlikely, since lowered GHG intensity generally corresponds to lower energy use and therefore lower expenses and higher profits); however, the expected growth in U.S. GDP during the same period was such that the reduction in GHG intensity would not reduce GHG emissions but instead only slow their growth.[8] Little activity resulted from this strategy. Meanwhile, major developments were occurring in the courts.

Court Action: *Massachusetts vs. EPA*

In 1998, the EPA legal office released an opinion that the Clean Air Act authorized the EPA to regulate GHGs. The next year a number of organizations petitioned the EPA to do so; specifically, they asked it to regulate emissions by motor vehicles of four GHGs: carbon dioxide, methane, nitrous oxide, and hydrofluorocarbons. The petitioners argued that greenhouse gases met the definition of "pollutant" prescribed by the Clean Air Act:

> Any physical, chemical, biological [or] radioactive substance or matter which is emitted into or otherwise enters ambient air [that] may reasonably be anticipated to endanger public health or welfare.[9]

By the time the EPA responded in August 2003, George W. Bush had become president. The EPA denied the petition, asserting that GHGs were not pollutants under the meaning of the Clean Air Act. In addition, they argued that the science on climate change was not clear, and that it was the role of the president and Congress, not the EPA, to determine climate policy for the United States.[10]

A number of states and advocacy organizations challenged the EPA's decision with a suit in federal court. In order to have such a suit considered, a plaintiff must have standing; that is, it must demonstrate that it is suffering actual harm from the policy it seeks to change. The state of Massachusetts had the strongest claim for standing to sue, namely that sea-level rise driven by climate change was damaging the state's coastline, a valuable economic resource; Massachusetts became the lead plaintiff in the case, which became known as *Massachusetts v. EPA*. The case was filed in the U.S. Court of Appeals for the District of Columbia, which has jurisdiction over most regulatory agencies. On September 13, 2005, the court ruled in favor of the EPA, but on June 26, 2006, the Supreme Court agreed to hear an appeal, with three issues to be considered: whether the plaintiffs had standing to sue, whether GHGs were "pollutants" according to the Clean Air Act, and, if they were pollutants, whether the EPA had the discretion to decide not to regulate them. On April 2, 2007, the court ruled in favor of the plaintiffs on all three issues and ordered the EPA to review GHGs as pollutants and determine if regulation was required.[11] By the time this review was completed, the United States once again had a new president with very different views on climate. The Supreme Court ruling provided President Obama with an important road toward climate action; before following that road, however, he tried some other directions.

Obama's First Term

President Obama took office with a very ambitious agenda. Internationally, he sought to restore America's status in the world while combating terrorism, promoting democracy, and ending the war in Iraq. In domestic policy, his goals included a stimulus bill to help bring the economy out of recession, enactment of a national health care system, action on climate, and reform of labor law, among other things. The stimulus and health care were clearly his top two priorities, but environmentalists had high hopes for climate action as well. Initially, at least, these hopes seemed to be disappointed, but such judgments turned out to be premature.

The Obama Campaign in 2008

Neither climate nor the environment in general was a major theme in Obama's successful campaign for president. Action to bring the country out of the recession and the country's troops out of Iraq understandably took precedence. Obama was also strongly committed to creating a national health care system. Nevertheless, he understood the need for action to reduce GHGs and control climate change, was clearly committed to protecting the environment, and appreciated the value of international cooperation. His directly coal-related campaign promises, as compiled by Politifact, included "create cap and trade system with interim goals to reduce global warming," "work with UN on climate change," and "create clean coal partnerships." Other goals relevant to climate included building a national high-speed rail network and launching a "green jobs" program. In other areas of environmental policy, Obama pledged to improve water quality and restore the Great Lakes.[12] After eight years in which George W. Bush had continued to express doubt about the strength of the scientific evidence for climate change, climate activists looked toward the new administration with optimism.

During his first two years, Obama did push consistently for climate action. In addition to his role at the Copenhagen UNFCCC Conference of the Parties in 2009, discussed in chapter 6, he asked Congress to pass a cap-and-trade program for GHGs and encouraged the EPA to begin the long process of reviewing CO_2 as a pollutant. However, he seemed to downplay climate issues after the midterm elections in 2010 gave control of Congress to the Republicans. Many environmental activists expressed dismay that the only environmental reference in his State of the Union address for 2011 was a joke about salmon. Others, such as Steven Cohen of the Earth Institute at Columbia University, pointed out that Obama had taken many actions to reduce GHGs that were not labeled as such, such as the support for solar electricity and high-speed rail in the stimulus bill, and suggested that he might be pursuing a stealth strategy.[13] This approach continued through the first two years of Obama's second term;

but in his final two years, he began to speak out more explicitly about climate, to refer to GHGs as "pollution," and to implement policies that did not require Congressional approval, notably his Clean Power Plan and the signing of the Paris Agreement under the UNFCCC.[14] Each of these phases of the Obama administration are examined more closely next, beginning with his push for a cap-and-trade bill during his first two years in office.

The Waxman-Markey Bill

From January 2009 until the January 2010 election of Republican Scott Brown to the Senate in a Massachusetts special election, the Democratic Party controlled the presidency and the House of Representatives, and held sixty seats, a "filibuster-proof" majority, in the Senate. In reality, it could never be assumed that all Democrats would vote the same way, so the supposedly filibuster-proof Senate majority could not be relied upon. Nevertheless, this period presented Obama with the best chance he would have to get important parts of his program enacted by Congress.

In previous Congresses, major climate bills had been cosponsored by Senators McCain and Lieberman; however, McCain had been Obama's Republican opponent in the 2008 election, and Lieberman had crossed party lines to endorse him. Coal mining interests had contributed $121,276 to McCain's campaign, compared to $19,950 to Obama's.[15] Whatever his reason, McCain declined to cosponsor a climate bill. Senators John Kerry (D-MA), Joseph Lieberman (I-CT), and Lindsey Graham (R-SC) tried to put together a bill but decided to wait for the House to act. In the House, the task of writing a bill and steering it through the chamber was assumed by Representatives Henry Waxman (D-CA) and Ed Markey (D-MA). Waxman was chair of the House Committee on Energy and Commerce, which had jurisdiction over air pollution, and Markey was the chair of that committee's Energy and Power Subcommittee.

In drafting the bill, Waxman and Markey worked closely with the U.S. Climate Action Partnership, a group of twenty-nine large corporations that supported limiting GHG emissions along with some large national environmental groups. The corporations, which included Duke Energy, an electric utility holding company, and Shell Oil, had concluded that emissions curbs were inevitable and wanted to play a role in designing them. The group spent $870,000 on lobbying in 2008, and $1 million in 2009.[16] The resulting bill prescribed GHG emissions reductions from the 2005 level of 3 percent by 2012, 17 percent by 2020, 42 percent by 2030, and 83 percent by 2050. Utility companies and others that emitted large amounts of CO_2 would be required to obtain federal emissions permits, which would set increasing lower levels each year in order to achieve the target reductions; 85 percent of the permits would be issued free, with the remainder sold by auction. The bill would also have required utility companies to get 20 percent of their energy from renewable sources by 2020, set national energy efficiency goals, and required the EPA to report to Congress on its progress. Following two days of intense lobbying by President Obama, the bill passed the House on June 26, 2009, by a vote of 219–212. Forty-four Democrats voted no, while eight Republicans voted yes.[17]

In the Senate, the House bill was referred to the Committee on Environment and Public Works, chaired by Barbara Boxer (D-CA). However, there did not seem to be enough votes to pass it due to the objections of some senators from states where manufacturing was important. In order to overcome this difficulty, Senators Kerry, Lieberman, and Graham decided to write a new bill that would approach different sectors differently. Electric utilities would be subject to a cap-and-trade system immediately, while manufacturers would have a different cap-and-trade regime that would not take effect for four years. The housing sector, a less important contributor to GHGs, would not face legal limits. However, in the summer of 2010, Majority Leader Harry Reid (D-NV) decided to defer action on the climate bill in order to give more priority to immigration, and Senator Graham, in turn, withdrew his cosponsorship in protest against Reid's action. The bill, which had never been assured of Senate passage, was effectively dead for the remainder of the 110th Congress. The election of a Republican House majority in the 2010 election made it unlikely that any bill regulating GHGs could be passed, and the administration began to look elsewhere for a means of action on climate change.

Stealth Approach

Following the death of the Waxman-Markey bill, President Obama seldom talked about climate and made no new climate proposals until his second term. However, he did present several climate-related programs under such headings as economic development and "green jobs." The American Recovery and Reinvestment Act of 2009, known more colloquially as the stimulus package, contained at least $105 billion for energy efficiency, renewable energy, and other environmental programs.[18] Other initiatives included having 1 million electric cars in operation by 2015 and obtaining 80 percent of the nation's energy needs from renewable sources by 2035.[19] These goals were linked explicitly to job creation in his

2011 State of the Union address, when he introduced them with the statement that "clean energy breakthroughs will only translate into clean
energy jobs if businesses know there will be a market for what they're
selling."[20]

The link of sustainability to jobs was a recurrent theme for Obama.
Soon after his inauguration, he had recruited Van Jones, an African American environmental activist, to join his team as a special advisor for green
jobs on the staff of the Council on Environmental Quality. Jones had published *The Green Collar Economy* in 2008 and had helped to establish the
Green Jobs Corps in Oakland earlier that year. As an activist and orga-
nizer, Jones was at first reluctant to accept a position in government, but
he ultimately agreed with the understanding that his job would be "to
help to shape policy to get as many jobs and as much justice as we can out
of the climate and energy proposals coming from the administration."[21]

Jones soon came under attack from Republicans for his youthful activities as a revolutionary and for inflammatory remarks about some Republican members of Congress. He resigned six months after his appointment,
but Obama continued to talk about renewable energy and other environmental goals as economic development and job creation. As Cohen put it,
"If the economy is the central issue of our political life, then environment
has just moved to that crucial central point. The 2011 State of the Union
address did not have to discuss climate policy to engage in it."[22] Obama
continued to stress the economy, rather than climate change, for the
remainder of his first term. Meanwhile, the EPA began to pave the way to
control GHGs through regulatory action under the Clean Air Act.

Implementing *Massachusetts v. EPA*

Shortly after the Supreme Court ordered the EPA to consider whether
GHGs should be regulated under the Clean Air Act, President Bush issued
Executive Order 13432 of May 14, 2007, which required the administrator of EPA, the Secretary of Transportation, and the Secretary of Energy to
work together in developing means of lowering GHGs from motor vehicles and stationary engines.[23] Then on December 5, 2007, the EPA sent
the Office of Management and Budget (OMB), which must approve all
proposed regulations, a proposed draft "finding that concentrations of six
key greenhouse gases in the atmosphere endanger the public welfare and
that emissions from new motor vehicles contribute to this problem." OMB
refused to consider this proposal, and the EPA then withdrew it; however,
they issued an advanced notice of proposed rulemaking (ANPR) on July
11, 2008. The notice included a technical supplement that reviewed the

existing scientific studies as well as responses to comments it had received to the earlier proposal.[24]

Soon after President Obama was inaugurated, the EPA took up the issue again and issued a new proposed finding, now including analysis of more recent studies and comments on the ANPR, on April 17, 2009. The EPA received over 380,000 public comments during the ensuing statutory sixty-day comment period; after reviewing these, and considering still more scientific studies, the EPA issued two findings on December 7, 2009. The 52-page document was accompanied by a 210-page technical support document, which surveyed the scientific literature, and five hundred pages (divided into eleven volumes) of responses to the public comments.[25] The first finding, known as the "Endangerment Finding," was "that the current and projected concentrations of the six key well-mixed greenhouse gases—carbon dioxide (CO_2), methane (CH_4), nitrous oxide (N_2O), hydrofluorocarbons (HFCs), perfluorocarbons (PFCs), and sulfur hexafluoride (SF_6)—in the atmosphere threaten the public health and welfare of current and future generations." In addition, a "Cause or Contribute Finding" was that "the combined emissions of these well-mixed greenhouse gases from new motor vehicles and new motor vehicle engines contribute to the greenhouse gas pollution which threatens public health and welfare."[26]

The Endangerment Finding was instantly criticized by conservative groups as a political attempt by the Obama administration to make an end run around Congress.[27] However, the finding was the result of a process that began with the filing of *Massachusetts v. EPA* in 1999 and was confirmed by the Supreme Court decision of 2007, and the drafting of the finding was a continuation of work begun by EPA during the Bush administration. While there is little doubt that the president encouraged the EPA to act and helped to steer the later issuing of regulations, the finding itself rests squarely on the application of current science to the letter of the Clean Air Act.

The EPA responded to the Cause or Contribute Finding about motor vehicle emissions by issuing ambitious new mileage requirements in 2011, but it took no further action to regulate GHGs until the president's second term. Meanwhile, two public campaigns to demand more climate action were helping to make the case for a tougher program.

Public Pressure for Obama to Act on Climate

Shortly after the disappointing end of the Copenhagen Conference of the Parties to the UNFCCC and the Kyoto Protocol, climate activists

Naomi Klein and Bill McKibben appeared in an online video interview,[28] looking as if they had seen the end of the world (in a written essay, McKibben remarks, "If you want to despair, that's certainly a plausible option")[29] and explaining everything that was wrong with the Copenhagen Accord. At the end, McKibben says that he does not know what the next step for the climate movement should be, but that "we'll figure it out." A year earlier, consultants Michael Shellenberger and Ted Nordhaus had proclaimed "the death of environmentalism."[30] A new approach was needed, and McKibben soon found it.[31]

The new approach began with science. James Hansen, the NASA scientist who had brought climate change into public awareness with his testimony to the Senate in 1988, had begun to shift his focus from reducing the annual rate of GHG emission to reducing the total amount of GHGs in the atmosphere. In a 2008 publication, Hansen and his team had analyzed paleoclimate data and determined that doubling the amount of atmospheric CO_2 from preindustrial levels would produce "a nearly ice-free planet, preceded by a period of chaotic change with continually changing shorelines." They concluded that "remaining fossil fuel reserves should not be exploited without a plan for retrieval and disposal of resulting atmospheric CO_2."[32] Adopting this approach, a 2013 report from the IPCC calculated that limiting additional atmospheric carbon to 900 gigatons would lead to a 33 percent probability that the temperature rise would be limited to two degrees (and therefore a 67 percent probability that it would be greater). A 50 percent probability could be obtained by holding additional carbon to 820 gigatons, and a 66 percent probability by holding it to 790 gigatons.[33] The panel added, somewhat grimly:

> A large fraction of anthropogenic climate change resulting from CO_2 emissions is irreversible on a multi-century to millennial time scale, except in the case of a large net removal of CO_2 from the atmosphere over a sustained period. Surface temperatures will remain approximately constant at elevated levels for many centuries after a complete cessation of net anthropogenic CO_2 emissions. Due to the long time scales of heat transfer from the ocean surface to depth, ocean warming will continue for centuries. Depending on the scenario, about 15 to 40% of emitted CO_2 will remain in the atmosphere longer than 1,000 years.[34]

Of course, the total amount of carbon added to the atmosphere is mathematically equivalent to the sum of the amounts added each year, so there is no substantive difference between a focus on emissions rates and a focus on total amounts of added carbon. Politically, however, the new approach to presenting the science led to new political strategy and tactics.

In the summer of 2012, McKibben published an article in *Rolling Stone* in which he adopted Hansen's findings on total atmospheric CO_2:

> Scientists estimate that humans can pour roughly 565 more gigatons of carbon dioxide into the atmosphere by midcentury and still have some reasonable hope of staying below two degrees. ("Reasonable," in this case, means four chances in five, or somewhat worse odds than playing Russian roulette with a six-shooter.)[35]

McKibben used this finding to advocate a strategy, namely, working to halt the development of any new fossil fuel reserves, and particularly of coal. This strategy had several advantages. Most important, it made it possible for climate activists to wage campaigns that were both negative and particular, and therefore unambiguous. It is useful to think of such campaigns in contrast with, for example, the campaign for Cape Wind, a large wind power development proposed for Nantucket Sound, off the coast of Massachusetts. Although most environmentalists supported the project, it was opposed by a coalition of wealthy Cape Cod residents (some of whom turned out to own coal mines) concerned about their view and environmentalists concerned about the unspoiled character of Nantucket Sound. Opponents of the project were always able to say that they were in favor of wind power, only not this particular project in this particular place. In the absence of any criteria, or any overall assessment of the need for wind power, this argument was not easy to counter.[36] Once the movement's goal was to stop burning coal and oil, rather than to develop the ideal energy system or to get the world to agree on who should reduce emissions by how much, organizing became simpler.

McKibben launched a new organization, 350.org, devoted to lowering the atmospheric concentration of CO_2 to 350 parts per million (ppm) and keeping it there. Current concentration is around 400 ppm, but 350 is believed by many scientists to be the highest level consistent with a global temperature rise of no more than 2°C. The organization would concentrate on mass mobilization and civil disobedience to pursue its ends. McKibben decided to focus on stopping the proposed Keystone XL (or KXL) pipeline, which would bring oil from the Alberta tar sands to refineries in the United States. The Keystone Pipeline, bringing tar sands oil in various forms from the extraction site to refineries in Oklahoma and Illinois, has been operating since June 2010.[37] An extension to bring oil from Oklahoma to the Gulf Coast in Texas is under construction.

TransCanada, the company that operates the pipeline, has proposed an additional extension—the KXL pipeline—that would leave the original pipeline in Alberta, run through an oil-producing region of Montana where it could pick up more oil, and end at the same terminus as the original pipeline in Oklahoma.

The focus on the KXL pipeline had several strategic advantages. First, the pipeline posed environmental risks that were locally important to the areas it would pass through, such as possible contamination of the Ogallala Aquifer in Nebraska, making it possible to broaden the anti-pipeline coalition beyond climate activists. A second advantage was that the amount of carbon present in the ground and available for extraction could be related directly to the amount that could be added to the atmosphere without causing irreversible damage. It was this relationship that led James Hansen to declare that it would be "game over for the planet" if all the oil sands oil was burned. Hansen explained that he chose to focus on the pipeline "because it's the first huge leap into unconventional fossil fuels. Once you set up that pipeline, it practically guarantees that a large fraction of that resource will be exploited."[38] McKibben, while agreeing with the other reasons for focusing on KXL, also cited political considerations: "One thing that made it attractive, from an organizing point of view, was that President Obama, himself, was charged with making this decision [it required a permit from the State Department in order to cross the international border]. His rhetoric on climate change was good, his performance less so."[39]

McKibben and 350.org launched a campaign to block the KXL pipeline in 2010. During the summer and fall of 2011, there were repeated acts of symbolic civil disobedience (e.g., sitting down on the sidewalk until arrested) in front of the White House, and a rally in November 2011 brought several thousand people to Washington to surround the White House with a human chain and a black plastic pipeline replica. In February 2013, 350.org joined with the Hip Hop Caucus and the Sierra Club to bring around 50,000 people to Washington for what McKibben called the biggest climate rally in U.S. history.[40] Opposition to the pipeline also led the Sierra Club to drop its long-standing refusal to join in civil disobedience.[41] Obama responded to the campaign by delaying a decision until after he was reelected, and he ultimately denied the permit.[42] Meanwhile, a second source of pressure on Obama was growing: the Sierra Club's Beyond Coal campaign.

In 2004 the Sierra Club, in partnership with several other organizations, launched the Beyond Coal campaign with a simple purpose: to oppose every proposal for a new coal-fired electric plant.[43] As of July 2010, the campaign had kept 132 proposed plants from being built, with others still in question, imposing what Lester Brown described as "a de facto moratorium on new coal-fired power plants."[44] The campaign then took on another goal, closing existing coal-fired plants. By the end of 2013, 158 plants, 20 percent of the total coal-powered generating capacity in the United States were scheduled to close, and no new plants had been begun for three years.[45] As of 2018, the total was 275 plants retired or proposed to be retired, with 255 to go.[46]

Local campaigns against particular power plants were frequently able to unite climate activists with people concerned about the destructive effects of coal burning on their neighborhood and with the campaigns against mountaintop removal in coal mining areas. While they faced the potential of opposition from unions representing workers in the power plants, they were often able to ally with these workers around economic development and retraining for jobs in green industry; as the perception spread that coal was in decline, such benefits became more attractive to miners.[47]

One activist described the anti-coal movement's tactics as "swarming"—using a combination of lawsuits, pressure on banks, community organizing, civil disobedience, and alliances with local small businesses, churches, and labor unions.[48] Legal action is important, and coal plants are vulnerable. Electricity is a regulated utility; power companies get a legal monopoly but give up their right to set prices to a regulatory agency, which in turn must set rates that let the companies cover their costs and earn a fair rate of return. Since costs are passed along to the consumer, the regulators are mandated to assure that electricity is generated in the least expensive way, and today that is not likely to be coal. The campaign was therefore able to challenge new coal plants on the grounds that they are too expensive, as well as contributing to sulfur pollution and climate change. It has done so enthusiastically; at one point the Sierra Club estimated that it was filing a lawsuit every ten days.[49] However, legal action is more effective when combined with a strong community voice—what one progressive foundation official refers to as "real people who are affected by these plants, who'll reject the economic argument publicly."[50]

The campaign has been much easier because of the economics of coal—or more important, the economics of natural gas. The development of hydraulic fracturing ("fracking") has made gas cheap and abundant. While it is true that both petroleum and natural gas emit less CO_2 per unit of electricity than coal, they still do contribute to climate change, and no environmentalist would consider them an acceptable substitute.[51]

For this reason, the Sierra Club has Beyond Oil and Beyond Natural Gas campaigns as well; gas is seen as only a step on the way to renewable energy, but in the short run the availability of natural gas makes it easier to target coal, a far dirtier fuel.

The Beyond Coal campaign's success in blocking and closing coal-fired electric plants made it possible for the Obama administration to use the Clean Air Act to regulate stationary-source carbon emissions; at a time when no new coal plants are being built, it was easier than it might have been to issue regulations to prevent their being built in the future. Obama began to move in this direction following his successful reelection campaign in 2012.

Obama Unbound: The Second Term

Barack Obama won a second term as president in 2012, with 61.7 percent of the popular vote and a 332–206 margin in the Electoral College. In the Senate his party gained two seats, for a total of fifty-three; since the two Independents caucus with the Democrats, its working majority was 55–45. In the House the Democrats gained eight seats for a total of 203, leaving the Republicans still in the majority by a margin of thirty-three seats. With the road to climate legislation still blocked by the House majority, and with no more need to maintain a broad personal electoral coalition, the president turned more boldly to actions he could take without Congressional approval. In his 2013 State of the Union address, Obama asserted his determination to act with or without Congress:

> I urge this Congress to get together, pursue a bipartisan, market-based solution to climate change, like the one John McCain and Joe Lieberman worked on together a few years ago. But if Congress won't act soon to protect future generations, I will. I will direct my Cabinet to come up with executive actions we can take, now and in the future, to reduce pollution, prepare our communities for the consequences of climate change, and speed the transition to more sustainable sources of energy.[52]

However, there would be one more Congressional election during Obama's presidency, and he continued to pursue a moderate course; two paragraphs after calling for executive action on climate, he declared, "My administration will keep cutting red tape and speeding up new oil and gas permits."

The midterm election did not go well for the Democrats; they lost nine Senate seats and their majority in that body, while the Republicans gained thirteen seats in the House for a majority of 347, their largest since 1928. With no more elections to lose, and less prospect than ever for getting his programs enacted by Congress, Obama and his administration turned more aggressively to executive action. Eight days after the 2014 election, the White House announced an agreement with China to reduce GHG emissions. The United States pledged a 26 percent reduction from 2005 by 2023 and to make its best efforts to achieve a 28 percent reduction, while China pledged to halt and reverse the increase in its GHG emissions by 2030, with its best efforts to achieve this goal earlier.[53] One year later, Secretary of State Kerry made an official decision that allowing the Keystone XL pipeline to cross the border was not in the national interest.[54] And in August 2015, the EPA issued the final version of its GHG emissions regulations for existing power plants, named the Clean Power Plan by the administration.[55]

The Turn to Executive Action

In the twenty-three years since the Republican congressional victories of 1994 there have been only slightly more than eight years of unified party government (i.e., situations where the same party simultaneously held the presidency and majorities in both the Senate and the House of Representatives). For the other seventeen years, the nation experienced what had become known as "divided government." Historically voters have seemed to prefer this condition, as whenever the presidency changed hands, control of one house of Congress soon changed in the opposite direction. Divided government was thought to be a way to check presidential power and to compel compromise on issues characterized by deeply held partisan beliefs.[56] More recently, however, divided government has tended to produce partisan gridlock rather than moderation and compromise as Congressional parties have put attempts to win back the presidency above the making of effective policy.[57] In response, presidents have looked for ways to make policy without Congressional consent. Barack Obama continued this trend. As Andrew Rudalevige has shown, Obama's unilateral presidential actions were not limited to formal executive orders but also included presidential memoranda, signing statements, statutory findings, letters, guidance documents, and administrative orders from department heads issued on orders from the White House.[58] Obama was to use several of these types of action in advancing his plan to control GHG emissions.

In addition to the actions described below, the president adopted new rhetoric. He continued to promote the economic and job-creation results of developing alternative energy, high-speed rail, and other climate-friendly projects. But in addition, he began to speak more directly of the threat of climate change, and to refer to GHGs as "pollution." This can be seen in his statement introducing the Clean Power Plan:

> Right now, our power plants are the source of about a third of America's carbon pollution. That's more pollution than our cars, our airplanes, and our homes generate combined. That pollution contributes to climate change, which degrades the air our kids breathe. But there have never been federal limits on the amount of carbon that power plants can dump into the air. Think about that. We limit the amount of toxic chemicals like mercury and sulfur and arsenic in our air or our water—and we're better off for it. But existing power plants can still dump unlimited amounts of harmful carbon pollution into the air.
>
> For the sake of our kids and the health and safety of all Americans, that has to change. For the sake of the planet, that has to change.
>
> So, two years ago, I directed Gina and the Environmental Protection Agency to take on this challenge. And today, after working with states and cities and power companies, the EPA is setting the first-ever nationwide standards to end the limitless dumping of carbon pollution from power plants.[59]

The word "pollution" appears five times in these three paragraphs. Years after the ruling in *Massachusetts v. EPA* and six years after the EPA's Endangerment Finding, Obama had begun to integrate that finding into the way he talked about GHGs and justified the regulations issued in order to control them under the Clean Air Act.

Applying the Clean Air Act

The original petition that led to the *Massachusetts v. EPA* ruling had asked the EPA to consider regulating emissions of four GHGs from motor vehicles. In consequence, the Endangerment Finding was accompanied by a finding that motor vehicle emissions of these gases did contribute to global warming (i.e., the Cause or Contribute Finding), and the first regulation pursuant to the finding was a strengthening of average fleet-mileage requirements for new vehicles. As with the Clean Power Plan, this regulation was challenged in court and then suspended by President Trump. Since it did not affect coal, we shall not consider it further here.

However, the Clean Air Act provides that if the EPA decides to regulate any substance in motor vehicle emissions as a pollutant, that substance must also be so regulated in emissions from stationary sources.

Specifically, it requires that any source emitting more than one hundred tons of a given pollutant must obtain a permit to do so. This numerical trigger, specified in the statute, poses a problem. While other pollutants result from incomplete combustion or from impurities in the coal burned, CO_2 is the main combustion product. In other words, all of the coal burned is converted into CO_2; as a result, the quantities produced are higher by several orders of magnitude. Around fifteen thousand point sources were required to have permits because of their emissions from traditional pollutants, but if the same standard was applied to emissions of CO_2, the number would rise to 6 million. It was estimated that the cost to the government of processing such permits would rise from $62 million to $21 billion, and the cost of compliance by the polluters would rise to $147 billion. These costs would have been staggering, and regulating emitters of such small amounts of CO_2 would have done little to reduce climate change. The EPA therefore issued a "tailoring rule" in 2010. Under this rule, sources would only require permits if they emitted more than 100,000 tons of CO_2 per year.[60]

Critics of the Clean Power Plan argued that the EPA lacked the legal power to set aside numerical criteria that were specified in the authorizing statute. The EPA responded that applying the statutory standard would produce an absurd result and that they were using administrative flexibility in order to attain the statute's intended results. This administrative overriding of statutory language became one of the legal issues by which the Clean Power Plan was challenged in court.[61]

In addition to regulating emissions, the Clean Power Plan addressed the question of ambient air quality. Once CO_2 was defined as a pollutant, each state had to include it in their State Implementation Plan (SIP) to bring air quality into compliance with the levels determined by the EPA. Doing so would require extending limits on CO_2 emissions to existing coal-fired utility plants as well as to new ones. Seeking to give states as much flexibility as possible, the EPA proposed to let states meet their emissions-reduction targets in a variety of other ways as well. For example, they might support new utility plants using alternative energy sources, such as solar and wind power, or take part in multi-state emissions-trading regimes.

This approach had some precedent. For example, SIPs to reduce ambient ozone might include such measures as limits on parking, support for car and van pools, and improvements in public transit. However, the EPA's proposal extended this approach into a new area, utility plant emissions, and became a focus of legal challenges. Twenty-nine states sued in federal court to block implementation of the Clean Power Plan. In January

2016 a three-judge panel of the U.S. Court of Appeals for the District of Columbia scheduled a hearing for June 2 of that year but refused unanimously to delay implementation of the regulations while the case was being heard. However, the Supreme Court reversed that decision by a vote of 5–4, ordering on February 9, 2016, the regulations be suspended pending final resolution of the case. The Supreme Court had never halted a regulation pending judicial review before; doing so in this case suggested that the five-judge majority was inclined to rule against the plan when the case reached them. The DC Court of Appeals finally heard the case on September 27, 2016, but it did not issue a ruling before the 2016 election and the announcement by President-Elect Trump that he intended to withdraw the plan, thereby rendering the issue moot.[62] Further legal action seems inevitable.

Conclusion

Obama's climate policy achievements are impressive. However, they may not be permanent. Faced with Congressional majorities determined to block action on climate, health care, and many other issues, the president found ways to act without Congress. The Paris Agreement was designed not to be a treaty so that it would not require Senate ratification; and the Clean Power Plan, along with the stricter motor vehicle mileage requirements, were imaginative uses of authority already delegated to the EPA by the Clean Air Act. With the election of Donald Trump, who during the campaign had asserted that climate change was a hoax propagated by China for economic advantage and declared his intention to end the "war on coal," the longevity of Obama's achievements was thrown into doubt. However, reversing existing policy may not be easy. Moreover, many state and local governments continue to pursue GHG reductions, and the other parties to the Paris Agreement are determined to carry on.

CHAPTER EIGHT

From President Trump to the Future

President Trump campaigned on a promise to end the war on coal and called climate change a hoax perpetrated by China to gain economic advantage. He promised to withdraw the United States from the Paris Agreement of 2015, to cancel the Obama-era Clean Power Plan, and to promote opening new coal mines and putting miners back to work. Environmentalists generally supported his opponent and viewed the coming Trump administration with dread. However, Trump was not able to go as far as he had wished in reversing Obama's climate policies, since there is considerable inertia in American public policy.

Major components of climate policy, such as the Clean Power Plan, are rooted in law. Although the plan can be revised, any new plan would have to be tested against the same statutory goals, and the latter could be changed only by the passage of new legislation. Similarly, while the president has the power to withdraw the United States from the Paris Agreement, the international structure for climate policy continues to grow and develop.

Finally, and perhaps surprisingly, the local government and private sector entities that had begun to pursue GHG reductions with the support of the Obama administration have mostly continued to pursue action without that support. Cities, states, and businesses throughout the United States have set GHG reduction goals for themselves and are pursuing those goals both individually and cooperatively through institutions such as the Regional Greenhouse Gas Initiative (RGGI), a cap-and-trade program involving ten states in the Northeastern United States, or ambitious

GHG reduction mandates of California's Global Warming Solutions Act of 2006.[1]

Climate in the 2016 Campaign

On November 6, 2012, Donald Trump posted on the social media forum Twitter, "The concept of global warming was created by and for the Chinese in order to make U.S. manufacturing non-competitive."[2] He maintained this position through most of his successful presidential campaign in 2016, although when Hillary Clinton cited his claim that it was a Chinese hoax in the September 26, 2016, presidential debate, Trump denied that he had said so. His campaign manager, Kellyanne Conway, explained the next day that he believed that some climate change was occurring but only from natural causes.[3] Most importantly, he affirmed repeatedly that he planned to end the war on coal and increase coal mining. After his election he did meet privately with reporters from the *New York Times* and stated there was "some connectivity" between human activity and climate change and that he had "an open mind" on the Paris Agreement;[4] however, this openness of mind did not seem be reflected in his subsequent actions.

If Trump's statements about coal were meant to get him votes, they seemed to succeed. He carried the coal states of West Virginia, Kentucky, Pennsylvania, and Ohio in the East and Wyoming and Montana in the West. Among the major coal states, only Illinois gave its votes to Clinton. Some of the coal states had usually gone Republican in recent elections, but Ohio and Pennsylvania were highly competitive; Clinton's losses there assured that she would lose the election.[5] Environmentalists looked to the incoming Trump administration with considerable trepidation.

Trump's First Two Years

President Trump mounted an all-out attack on his predecessor's climate policy during his first two years in office, beginning just a few weeks after his inauguration. Very early on, Vice President Mike Pence received a memo from coal magnate Robert Murray, the head of the Murray Energy Corporation, who had contributed $300,000 to the inauguration celebration. The memo contained what Murray called "Action Plan" to promote coal. The plan was not limited to climate issues. In addition to repealing the EPA's Endangerment Finding on GHGs, withdrawal of the Clean Power Plan without a replacement, and withdrawal from the Paris Agreement, Murray also proposed weakening the protections for miners from coal dust and black lung disease and replacing the boards of the Federal Energy Regulatory Commission, Tennessee Valley Authority, and National Labor Relations Board. Contrary to the expressed view of the coal industry, Murray also recommended defunding CCS technology, which he referred to as "a pseudonym for 'no coal.'" The plan also called for ending tax credits for wind and solar power.[6] Much of Trump's coal agenda corresponded with the points in this memo.

The president appointed people to key environmental positions who denied the reality of the greenhouse effect, withdrew the United States from the Paris Agreement, and sought to replace the Clean Power Plan with an alternative seen by most as ineffective. He also sought large reductions in both the number of staff and the budget of the EPA, thereby making any enforcement actions difficult. Nevertheless, there are structural limits on what any president can do unilaterally, so the status of many of these actions remains uncertain. Each of these actions is evaluated in more detail in this section.

Trump's Appointments

President Trump's first nominee to head the EPA was Scott Pruitt. Pruitt had served as the attorney general of Oklahoma, an elected position, from 2010 until his appointment to the EPA; in that capacity he had sued the EPA on fourteen occasions, objecting to the Clean Power Plan, the Cross-State Air Pollution Rule, and the Waters of the United States (WOTUS) Rule, among others. He also sued the EPA for settlements it had agreed to in lawsuits filed by environmental groups.[7] None of the suits were successful. Pruitt had also stated his disbelief in the idea that GHG emissions were causing climate change.[8] He later suggested that a rise in global temperature might prove to be a good thing, asserting, "That's fairly arrogant for us to think that we know exactly what it [i.e., global surface temperature] should be in 2100."[9] Pruitt was confirmed, 52–46, on February 18, 2017, with the vote largely on party lines (two Democrats voted to confirm, while one Republican voted against.)[10]

Pruitt's time in office was marked by scandals involving lavish expenditures, use of government funds and EPA staff for personal tasks (including helping his wife find a high-paying job), and renting a condominium at far below market rates from the wife of J. Steven Hart, a lobbyist for Exxon, Cheniere Energy, and other oil companies regulated by the EPA. He spent $100,000 on first class air travel and another $43,000 on travel in private planes, as well as $45,000 on construction of a soundproof booth in his office. He insisted on twenty-four-hour security from a twenty-person detail, including on family trips to the Rose Bowl and Disneyland, at a total cost of over $3 million. He was also charged with evading public records laws by using four different email accounts and making official phone calls from other people's phones. Under investigation for these and other charges by the inspector general of the EPA and the House Oversight Committee and facing a lawsuit from Public Employees for Environmental Responsibility over his recordkeeping, Pruitt resigned on July 5, 2018.[11]

During his tenure, Pruitt initiated the withdrawal of several important environmental regulations. In addition to the Clean Power Plan (discussed further later), these withdrawals included higher fuel efficiency requirements for automobiles and the WOTUS rule that extended federal protection to more of the nation's wetlands. However, none of these actions had been completed by the time of his resignation. The Clean Power Plan and the WOTUS rule, which had already been halted by litigation, remained suspended, but otherwise all environmental regulations that had been in force when Trump took office were still in force when Pruitt resigned.

Following Pruitt's resignation, his deputy, Andrew Wheeler, became acting administrator of the EPA. Wheeler had previously been a lobbyist focusing on energy issues and had worked for Robert E. Murray, head of Murray Energy Corporation and the author of the pro-coal action plan discussed earlier. In an interview with Lisa Friedman of the *New York Times*, Wheeler said that he had worked with Murray to block the Clean Power Plan but that he had not been involved with the action plan memo; he also said he had ceased to lobby the EPA after Trump's election because he anticipated being appointed to that agency and wanted to avoid violation of the ethics rules.[12] On November 16, 2018, the president announced his intention to appoint Wheeler as administrator.[13]

Trump's choice for Secretary of the Interior, Rep. Ryan Zinke (R-MT), was equally repugnant to environmentalists. Although he had once called on the Obama administration to take strong action to prevent climate change, and had stated that he had personally observed glaciers receding in Glacier National Park, Zinke had later expressed doubt that the change was the result of human action. In any case, his changing views on climate had never interfered with his advocacy of greater use of coal, possibly combined with CCS, to meet America's energy needs. He had supported the Keystone XL pipeline and condemned government action to protect the sage grouse. Citing former President Theodore Roosevelt, Zinke argued, "Conservation means development as much as it does protection."[14] In a 2016 book about his military experiences, Zinke had complained that it was unfair for oil companies to have been fined for "the alleged death of a handful of ducks" after an oil spill.[15] Unlike many conservative Republicans, Zinke has opposed a wholesale transfer of federal lands to the states and resigned his position as a delegate to the 2016 Republican National Convention over this issue. However, his view that federal lands should be better managed rather than transferred seems to mean that the federal government should be more active in opening the lands to oil, gas, coal mining, and other development.[16]

Like Pruitt, Zinke was criticized for extravagant travel expenditures, particularly the use of charter planes. As of early October 2018, the Interior Department's inspector general had opened fourteen investigations into his conduct; on October 15, Zinke fired the inspector general, Mary Kendall.[17] In November 2018, a probe over the sale of land in Whitefish, Montana, owned by Zinke and his wife to the chair of the Halliburton Corporation, which has multiple business dealings with the Interior Department regarding its oil-drilling business, had been turned over to the Department of Justice for possible criminal prosecution.[18] When Rep. Raul Grijalva (D-AZ), incoming Chair of the House Committee on Natural Resources, called on Zinke to resign, Zinke responded by calling Grijalva a drunk.[19] Zinke resigned from office on December 15, 2018.[20]

President Trump also made a concerted effort to appoint conservative judges to federal courts, including the appointments of Neil Gorsuch and Brett Kavanaugh to the Supreme Court. He has worked closely with the Federalist Society, a group of conservative lawyers, and its executive vice president, Leonard Leo, to choose nominees. These appointments are important in the long run because they may reduce the ability of environmentalists to block executive actions that are not in accord with the Clean Air Act, a matter that will be discussed later in this chapter. Such appointments are an ongoing drive by Trump and the Senate Republicans.[21]

In addition to his appointments, Trump took several dramatic actions designed to promote coal, beginning with his withdrawal from the Paris Agreement on climate change.

Withdrawing from the Paris Agreement

The Paris Agreement under the United Nations Framework Convention on Climate Change (UNFCCC), discussed more fully in chapter 6, was widely hailed as a major step toward controlling GHG emissions as well as a diplomatic triumph for Barack Obama. As with many Obama policies, Donald Trump campaigned for president with a promise to reverse this one by withdrawing the United States from the agreement. He kept this promise, announcing U.S. withdrawal on June 1, 2017, and declaring, "I was elected to represent the citizens of Pittsburgh, not Paris."[22] He stated specifically that the United States was cancelling its Intended Nationally Determined Contribution to GHG reduction, which would have meant a reduction of 26–28 percent below 2005 levels by 2025, and would not fulfill its pledge of $3 billion to the Green Climate Fund, a UN fund that helps poorer countries adopt less-polluting technology.

According to press reports, a team of Trump insiders, including his daughter Ivanka, his son-in-law Jared Kushner, and then Secretary of State Rex Tillerson, had sought to persuade him to change the U.S. Nationally Determined Contribution to a lower amount while remaining a party to the agreement, while then EPA administrator Scott Pruitt had argued for withdrawal; Trump chose to withdraw. He may have wanted to keep the campaign promise as visibly as possible. However, he also sought to keep his options open; even while announcing the withdrawal, he added, "We are getting out. But we will start to negotiate, and we will see if we can make a deal that's fair. And if we can, that's great."[23] Under the rules of the Paris Agreement, the U.S. withdrawal will not take effect until November 2020, and there have been repeated hints from the administration that the United States might decide to stay in after all. Early in 2018, George David Banks, who had recently resigned as the White House senior advisor on energy and climate change, said in an interview, "The Paris agreement is a good Republican agreement. It's everything the Bush administration wanted." He added, "A lot can happen between now and 2020" and "we could conceivably go back in."[24]

Despite its announced intention to withdraw from the agreement, the U.S. delegation played an active role in the 2018 Conference of the Parties in Katowice, Poland (COP 24). The U.S. role at this conference was characterized by one observer as "somewhat schizophrenic;" its delegation helped block endorsement by the conference of the warnings in the October 2018 special report of the Intergovernmental Panel on Climate Change (IPCC) that the global temperature was rising faster than expected,[25] but it also lobbied hard for the adoption of strict and transparent measures of GHG emissions that would apply to every member country, winning significant concessions from China on this point.[26]

For the most part, with the exception of ceasing its contributions to the Green Climate Fund, Trump's announcement of withdrawal from the Paris Agreement is largely symbolic; its real effect hinges on the outcome of the 2020 presidential election. The ambitious goals of the INDC have been canceled, but the INDC itself was not legally binding; it was simply a list of actions and events that the United States planned or expected.

The real struggle will be over the fate of those planned actions, of which one of the most important is the implementation of the Clean Power Plan.

Repealing the Clean Power Plan

On October 9, 2017, EPA Administrator Scott Pruitt announced that he had signed an order to repeal the Clean Power Plan. The order would repeal the plan in its entirety, without replacement; however, most power companies argued that it should be replaced by a new plan for two reasons. First, the 2009 Endangerment Finding was still in effect, so the EPA was legally required to do something to reduce GHG emissions. Moreover, if no plan at all were in place, a future administration less friendly to coal would be free to write its own, without first having to go through the lengthy process of repealing the existing regulations.[27] Pruitt reportedly did not accept these arguments at first, preferring to challenge the scientific basis of the Endangerment Finding itself, as had been recommended in Murray's Action Plan; however, Pruitt eventually agreed. On December 27, 2017, the EPA submitted an Advance Notice of Proposed Rulemaking, asking for input on a replacement rule to limit GHG emissions from existing electrical generating plants. The thirteen-page document did not suggest particular rules but gave examples that suggested that they intended to stress technical improvements such as more efficient burners, rather than alternative energy sources, and that they sought suggestions for state guidelines, rather than federal regulations.[28]

Although the proposed repeal and the proposed replacement were logically related, they proceeded along separate legal tracks. The fifteen-page proposal to repeal the Clean Power Plan focused on the legal argument that the latter had exceeded the authority given to the EPA under the Clean Air Act. In essence, the EPA was accepting the argument of the twenty-seven states that had previously sued it, an argument summed up as follows:

> In other words, as applied to both new source standards and existing source standards promulgated under CAA section 111, if standards must be set for individual sources, it is reasonable to expect that such standards would be predicated on measures that can be applied to or at those same individual sources.[29]

The provisions of the Clean Power Plan that called for emissions trading or replacing coal power with alternative energy sources as a way to meet required GHG reduction were therefore rejected.

The EPA seemed not to have anticipated the volume of comments they would receive. The initial proposal had stated that a public hearing would be held if anyone requested it; the next day the EPA announced that there would be two days of public hearings, November 28 and 29, 2017, in Charleston, West Virginia, and the period for written comments would be kept open until January 16, 2018. On February 1, 2018, they announced that the public comments period would be reopened until April 26, 2018, and that there would be three "listening sessions" for further oral comments to be held on February 21 in Kansas City, Missouri, February 28 in San Francisco, California, and March 27 in Gillette, Wyoming. Comments did close on April 26; as of yet there has been no further action.[30]

While action was pending on the proposed withdrawal, the EPA submitted a proposed new rule, the Affordable Clean Energy (ACE) Rule, on August 21, 2018. The proposed rule would require states to set their own standards for CO_2 emissions from power plants within the state, and it was expected to reduce GHG emissions by 0.75 to 1.5 percent, in contrast to the 19 percent expected if the Clean Power Plan had gone into effect. States would be limited to changing the emissions levels for individual plants, with none of the trading or offset options permitted by Obama's plan.[31] As of November 2018, public comment had closed for both the repeal of the Clean Power Plan and the new ACE rule, but neither had been issued in final form.

It is not unusual for environmental regulations to take a long time to reach final form. The law requires the EPA to review existing scientific findings and to take into consideration all public comments that were submitted in response to a draft proposal. The scientific review is likely to be particularly difficult in this case since the decision was made not to withdraw the finding that GHG emissions posed a danger to human health and the environment; in consequence, any new plan that does not protect human health will be vulnerable to legal challenge. Ironically, the EPA's task in writing the final regulation is likely to be more difficult because of the drastic cuts in budget and personnel that were imposed on it by former administrator Scott Pruitt. These cuts will be examined in the next section.

Weakening the EPA

The Trump administration's first budget proposal for the 2018 fiscal year included a 30 percent cut in funds for the EPA, the largest cut for any agency. The cuts were attacked by members of both parties when Administrator Pruitt appeared before the relevant subcommittee of the House

Appropriations Committee on June 15, 2017. The subcommittee chair, Rep. Ken Calvert (R-CA), declared that "the budget proposes to significantly reduce or terminate programs that are vitally important to each member on this subcommittee." The subcommittee's ranking minority member, Rep. Betty McCollum (D-MN), said, "The budget that you have come before us today to support would endanger the health of millions of Americans, jeopardize the quality of our air and water and wreak havoc on our economy." Rep. Nita Lowey, the ranking minority member of the full Appropriations Committee, linked the proposal to other criticisms of Pruitt:

> Between your disturbingly close ties to the oil and gas industries, your past work to directly undermine the EPA and your skepticism that human activity plays a role in climate change, I suppose it's surprising you didn't propose to eliminate the agency altogether.[32]

In July, Representative Calvert released the draft of a subcommittee bill that would have reduced Trump's proposed $2.6 billion cut in the EPA budget to $528 million.[33] By the time the appropriations bill had made its way through Congress, the cuts to the EPA had been fully restored; President Trump signed the bill appropriating $8.6 billion for the EPA into law on March 27, 2018.[34]

Even as his proposed cuts to the EPA budget for fiscal 2018 were being rejected, President Trump proposed a cut of $6.1 billion, a 26 percent reduction, for fiscal 2019. The proposed cuts would involve a 21 percent cut in EPA staff. Once again, both the House and the Senate insisted on restoring the cuts. As of December 2018, a final budget had not been agreed on by Congress, but the differences between the House and Senate primarily involved the inclusion of policy provisions in the bill; the proposed amounts were very close, $35.3 billion in the House and $35.8 billion in the Senate for the combined Interior Department and EPA budgets, which are traditionally included in the same bill. However, the House had included a large number of policy provisions that the Senate had omitted, most notably a Congressional repeal of the controversial WOTUS regulation issued during the Obama administration. As a result, the conference report had not reached the House and Senate floors by the end of 2018.[35]

A second effort of Trump and Pruitt was to reduce the number of EPA personnel. In early 2017, they announced plans to cut the agency's workforce by twelve hundred jobs by September 2018, and they launched a buyout program to achieve this. This effort was more successful than the budget cuts. As of September 2017, EPA employment was about six hundred less than the year before, a total of 15,058; by June 2018 the number had been reduced further to 14,580. Meanwhile, Secretary Zinke announced plans to cut Department of the Interior staff by four thousand over one year.[36].

To sum up, President Trump and his administration have initiated the reversal of many policies promulgated under his predecessor in support of his proclaimed goal of increasing coal production and bringing back mining jobs. As of this writing, none of these initiatives has been adopted into regulations. With one house of Congress about to pass into control of the Democratic Party, which generally favors strong climate policy, it is unlikely that Trump will be able to enact any of his proposals through legislation. He must therefore convince the federal courts to overturn precedent and allow the implementation of new regulations that seem, on their face, to violate the language of the Clean Air Act. Trump's hope is that if he can appoint enough new conservative federal judges, the courts may begin to change their interpretation of the law. Environmental advocates hope, on the other hand, that they can delay significant damage to both the structure of environmental law and the environment itself until the next election, when they hope to see Democratic victories for the president and the Senate majority—neither of which is assured.

Although the policy of the U.S. government on coal and climate is tremendously important, it is not the whole story. Even as the Trump administration seeks to revive coal, many state and local governments remain committed to GHG reductions, as do many businesses; and the remaining members of the Paris Agreement continue to seek a path toward holding the global temperature rise below two degrees Celsius. Finally, grassroots activists are having some success in the campaign to move the United States and the world beyond coal. The remainder of this chapter will look briefly at each of these developments.

State and Local Action

Even as the Trump administration sought to cancel policies to reduce GHG emissions, many state and local governments pledged to move forward on their own. Their reasons for doing so are complex. Logically, the actions of a single city or state could not have a measurable impact on the climate, but many local elected officials seem to want to be seen as enlightened and progressive leaders, and urban voters seem to value such a stance. In addition, many of the policies that reduce GHG emissions also improve the quality of everyday life by providing safe bicycle lanes, better mass transit, and cleaner air. Calculating the total impact of all state and local action is beyond the scope of this book, but the following sections discuss some representative cases by city, state, and regional entities.

Mayors and Cities

Chapter 6 described the signing of the international Compact of Mayors in conjunction with the Paris Agreement in 2015. The compact joined with the EU-based Covenant of Mayors to form the Global Covenant of Mayors for Climate and Energy (GCoM) on June 22, 2016. As of January 2019, it had 9,258 local governments in its membership. GCoM provides a framework for recording and monitoring commitments by local governments to GHG reduction, clean energy production, and other climate action. As of September 2018, it had received 9,149 commitments that would total 1.4 billion tons per year of CO_2 emissions reductions by 2030, the equivalent of all expected U.S. automobile emissions in that year.[37]

Within the United States, the then mayor of Seattle, Greg Nickels, launched a Mayors Climate Protection Agreement on February 16, 2005—the day the Kyoto Protocol went into force. Nickels had sought for the agreement to be signed by 141 cities, a number chosen as equal to the number of member states in the Kyoto Protocol, a goal achieved by the annual meeting of the U.S. Conference of Mayors (USCM) later that year. The compact gained its 500th signatory (Tulsa, Oklahoma) by May 2007 and has 1,060 in early 2019. Signatories pledge to "strive to meet or beat" the goal set for the United States by the Kyoto Protocol (i.e., 7 percent GHG reduction from 1990 levels by 2012), to urge their state and federal governments to do the same, and to support passage of a bipartisan GHG reduction bill by Congress.[38] Although the year 2012 has come and gone, the specific target in the mayors' agreement has not been revised; nevertheless cities continue to set their own goals for GHG reduction, with thirty-five exemplar cases described in a USCM report of 2015.[39]

Several individual cities have adopted ambitious climate agendas. Globally one of the earliest to do so was London under the leadership of its first mayor, Ken Livingstone.[40] Livingstone's government enacted a "congestion charge" that had to be paid daily by any automobile driving into an eight-square-mile area in central London, while simultaneously improving bus transit, policies designed to both ease traffic congestion and reduce GHG emissions from motor vehicles. In 2007, he released a detailed plan to cut carbon emissions by 60 percent over twenty years.[41] Livingstone, who was elected as an Independent in 2000 and reelected as the Labour candidate in 2004, was defeated in 2008 by the conservative Boris Johnson. Johnson sought to reverse several of Livingstone's policies but left both the congestion charge and the commitment to 60 percent reduction in GHGs by 2025 in place.[42]

A number of U.S. cities have sought to follow London's example. For example, the city of Boston has set goals for reduction in GHGs from municipal operations of 25 percent below 2005 levels by 2020, and 80 percent reduction by 2050. The plan, named Greenovate Boston, also includes preparation for adapting to climate change, particularly sea-level rise, and aims to develop a cooperative approach among various levels of government, private business, and city residents. The eighty-page plan makes specific recommendations in four focus areas: neighborhoods, large buildings, transportation, and climate preparation. The plan concludes with a vision for "80×50," which is meant to "serve as a guide for the development of integrated policies and projects across public and private sectors." The city pledges to "ensure that this long-term goal is included in transportation, housing, and other planning efforts so that this transformation is embedded into all aspects of Boston life."[43] Responsibility for the plan has been assigned to the city's Environment Department, with line items in the annual city budget, a total of $2.3 million in fiscal year 2019, together with $5 million to study the feasibility of building a barrier against sea-level rise in Boston Harbor.[44]

The city of Chicago was an early adopter of city climate plans. Mayor Richard Daley appointed a task force to draw up a plan in 2006. The task force worked with business leaders, nonprofits, environmentalists, city departments, neighborhood activists, and others, ultimately releasing a sixty-page plan in 2008. The Chicago Climate Action Plan (CCAP) proposed to reduce GHG emissions by 80 percent below 1990 levels by 2050, with an interim goal of 25 percent reduction by 2020. This reduction was to be achieved in four strategic areas: energy efficient buildings (30 percent), clean and renewable energy sources (34 percent), improved transportation options (23 percent), and reduced waste and industrial pollution (13 percent). Each strategic option was further broken down into specific planned actions, such as "making appliances work for us" under energy efficient buildings. There were thirty-five actions in total, together with thirty-five proposals for adapting to climate change.[45]

The plan was a conceived as a strategic partnership between Chicago city government and the nonprofit Global Philanthropy Partnership (GPP), an arrangement meant to broaden the focus beyond action by municipal agencies to what every business and resident in the city could contribute. Two city officials and the chair of the GPP provided collaborative leadership for the plan, while the budget drew on both city revenues and grants from foundations.[46] In a comparative study of climate action plans in selected U.S. cities, Hugh Bartling concludes that "Chicago's experience, thus far, can probably be considered a model for planning to mitigate greenhouse gas emissions at the local level that thrived on multi-faceted leadership practices," but he added that this was due in part to local circumstances (such as existing technical networks and the presence of strong community philanthropies in the city) that might not be replicable elsewhere.[47] One of those circumstances was the strong commitment to the plan of Mayor Daley. When Daley left office in 2011, his successor, Rahm Emanuel, discontinued the CCAP in favor of his own initiative, Sustainable Chicago, which combines climate change with a number of other environmental concerns. Goal 22 of Sustainable Chicago is to "reduce carbon emissions from all sectors," but no quantitative targets are specified.[48]

While the transition from the CCAP to Sustainable Chicago 2015 seems like a step backward, the city and its mayor remain committed to climate action. In a 2017 statement opposing President Trump's repeal of the Clean Power Plan, Mayor Emanuel said that "Chicago is committed to taking up the reins of leadership on climate change" and that it had "reduced its carbon emissions by 11 percent from 2005 to 2015, bringing the city to 40 percent of the way to meeting its Paris Climate Agreement goals."[49]

Cities are taking significant actions to reduce GHG emissions. However, such actions are not systematic, nor is their continuation certain in the face of political change. Their actions offer hope but need to be supported by action elsewhere. One source of further action is the U.S. states, which we shall look at next.

States

California has a population of about 40 million people; were it an independent country, its economy would be the fifth largest in the world.[50] The state's size, along with a tradition of progressive government and a historic concern with air pollution, have given the state the political will to undertake a strong climate action plan. In 2006, the California legislature passed Assembly Bill 32, requiring a reduction in GHG emissions to 1990 levels by 2020; the responsibility for implementation was assigned to the state's air pollution control agency, the California Air Resources Board (CARB), supported by a Climate Action Task Force of other relevant state agencies. The bill required the drawing up of a Climate Scoping Plan every five years to spell out how the mandated reductions were to be achieved.[51] A new law, requiring GHG reductions of a further 40 percent by 2030, went into effect in 2017. The state has operated a cap-and-trade system for GHG reductions since 2012 and actively promotes zero-emissions vehicles and the development of wind, solar, and other clean energy sources. CARB reports that the state is on track to meet both the 2020 and 2030 goals.[52] California's commitment to its GHG reduction goals has persisted through both Republican and Democratic governors.

Environmental justice groups have challenged the law on both procedural and substantive grounds. Procedurally, they have argued that the participation of communities of color in policy decisions, although mandated by the law, has been tokenistic. Substantively, the challengers have argued that the cap-and-trade system administered by CARB results in the concentration of pollution in communities of color and other low-income areas, which have weaker political ability to resist such concentration. CARB has focused on reducing CO_2 emissions, and since CO_2 is not harmful to people, animals, or plants, CARB has not been concerned about local concentrations. However, facilities that emit CO_2 tend to emit more hazardous pollutants, such as sulfur and nitrogen oxides and particulates, as well. In June 2009 a coalition of environmental justice groups sued CARB over these issues. The court ordered CARB to produce a more thorough analysis of the environmental justice impacts of the program, which CARB did. However, few substantive changes were made, and the court then let the program go forward.[53]

Few states except California are large enough to tackle GHG reductions on their own, although many have adopted such policies as requiring utilities to increase the share of electricity produced from renewable fuels and providing financial assistance to individuals and small businesses for insulation and decentralized solar power. To gain the advantages of scale, several states in the Northeast have entered into an interstate agreement, the Regional Greenhouse Gas Initiative (RGGI), which will be described in the next section.

Regional Greenhouse Gas Initiative

On December 20, 2005, the governors of Maine, New Hampshire, Vermont, Connecticut, New York, New Jersey, and Delaware signed a memorandum of understanding to create the RGGI, a mechanism to reduce GHG emissions from power plants. Rhode Island and Maryland joined the agreement in 2007. New Jersey withdrew from the agreement under Republican Governor Chris Christie as of January 1, 2012; Christie's Democratic successor, Phil Murphy, signed an executive order to return soon after he took office, a proposal currently under consideration by RGGI. Virginia is also seeking to join the initiative.[54]

RGGI is governed by a board of directors comprised of the heads of environmental and energy regulatory agencies in each member state. The agreement set an initial regional cap of 188 million tons of carbon emitted in 2008 from electric generators of twenty-five megawatts or higher capacity powered by fossil fuels. The cap was lowered to 165 million tons after New Jersey withdrew and is reduced by 2.5 percent each year through 2020. In August 2017 the members agreed to an additional 30 percent reduction by 2030, at which time emissions from the current nine members would be capped at 55 million tons.[55] The total amount of the cap is allocated to the member states, each of which auctions emissions permits to the plants within the state. The money raised by the auctions is used to pay for renewable energy and energy conservation programs. RGGI is the first cap-and-trade program to sell permits at auction, rather than distributing them for free based on past usage; doing so increases the value of conserving energy while avoiding the criticism that earlier programs created windfall profits for the polluters.[56]

RGGI has been very successful. The compliance rate among power plants has been 96 percent, so the region is on track to achieve the planned 50 percent reduction in GHG emissions from power plants by 2030. At the same time, the auctions have generated hundreds of millions of dollars in clean energy investments. It is not a complete solution, as it does not cover such major sources of GHGs as motor vehicles, manufacturing, agriculture, and buildings. However, it does offer a model that can be applied in other areas, as well as showing the potential for states to cooperate on climate change without relying on the federal government. Meanwhile, developments in the international arena are showing what can be done in the absence of U.S. government participation.

The Future of International Climate Action

The partisan character of climate politics in the United States carries over to the role of the United States in international climate negotiations. The result has been a series of policy reversals. The first President Bush, a Republican, signed the UNFCCC only after all quantitative goals for GHG reduction had been removed; his Democratic successor, Bill Clinton, sent Vice President Gore to Kyoto to push for a binding agreement. The next Republican, the second George Bush, withdrew the United States from the Kyoto Protocol; his Democratic successor, Barack Obama, worked hard to secure international consensus on the Paris Agreement, winning major commitments from China and adopting meaningful GHG reductions in the United States. Now Obama's Republican successor, Donald Trump, has withdrawn the United States from the Paris Agreement. Other nations, with only a few exceptions, have stuck to their climate policy regardless of politics. In general, these nations have chosen to carry on without the United States at those times when the United States has withdrawn from an agreement, seeking to make some gains in GHG reduction while recognizing that in the long run the participation of the United States, like that of China, will be essential to success.

The initial reaction of the world community to the U.S. withdrawal from the Paris Agreement was to assert that the agreement would continue. Donald Tusk, the president of the European Council, declared that the effort to restrict climate change "will continue with or without the United States." German Chancellor Angela Merkel said that the U.S. withdrawal "will not deter all of us who feel obliged to protect this earth," and Koichi Yamamoto, Environment Minister of Japan, said that Trump "turned his back on the wisdom of human beings." Hua Chunying, spokesperson for China's Ministry of Foreign Affairs, said "China is willing to enhance cooperation with all sides to together advance the follow-up negotiations on details of implementing the Paris Agreement and also advance its effective implementation."[57] French president Emmanuel Macron, directing his remarks in English to the American public, announced, "To all scientists, engineers, entrepreneurs, responsible citizens who were disappointed by the decision of the president of the United States, I want to say that they will find in France a second home."[58]

The formal process of the Paris Agreement has continued to move ahead. Annual Conferences of the Parties were held in Bonn in November 2017 and in Katowice in December 2018, with meetings of subsidiary bodies in Bonn and Bangkok. The Katowice Conference achieved agreement on how each nation's GHG reductions would be measured and on transparency in reporting those reductions; it was considered a significant step forward. The United States, which is still a member of the Paris Agreement until its announced withdrawal takes effect in November 2020, signed the agreement reached in Katowice, a step that will make it easier should a new president decide to return to the agreement in the future.[59]

The United States is the world's second-greatest carbon polluter. Ultimately, climate change cannot be controlled without U.S. participation. But the international structures to control GHGs are in place; Trump's withdrawal, while unfortunate for the climate, could be reversed by a political change in the future. Meanwhile, there is one more force for GHG reduction: the organized action of private individuals. Citizens groups such as the Sierra Club, 350.org, MoveOn.org, Greenpeace, Clean Water Action, and others have long been campaigning for stronger climate policy. Since the 2018 election, which brought a Democratic Party majority to the U.S. House of Representatives, these movement groups have begun to find a focal point within the Democrats in Congress, centered on the concept of a Green New Deal.

Grassroots Climate Action and the Green New Deal Proposal

Chapter 7 discussed the Sierra Club's Beyond Coal campaign to close coal-fired power plants. That campaign has continued; as of February 2019, the number of coal plants retired or proposed for retirement stood at 282, with 248 left to go.[60] Despite President Trump's declared effort to revive coal power, plants with a capacity of 13.7 gigawatts, a record, were closed in 2018. However, U.S. CO_2 emissions actually increased by 3.4 percent that year, reversing three years of decline. Despite increases in renewable power, both replacement of the capacity of the closed plants and the energy needed to meet increased demand came mostly from new natural gas plants. While natural gas is less polluting than coal, it still produces GHGs. These combined with increased emissions from the transportation sector, from buildings, and from industry to produce the overall increase.[61]

The United States may have reached a point where real gains in preventing climate change will require strong policy at the federal level. It is appropriate, therefore, that at the beginning of 2019, grassroots climate activism has begun to focus on the concept of a Green New Deal. This concept is intended to evoke Franklin Roosevelt's program of economic stimulus and direct job creation policies in both its massive scale and its promotion of greater economic equality, while adding a focus on fighting climate change and creating a sustainable, prosperous economy. The concept has become particularly associated with the newly elected Rep. Alexandria Ocasio-Cortez (D-NY) who has worked with movement groups to pressure other members of Congress to support it.[62]

Ocasio-Cortez, a twenty-nine-year-old activist from the Bronx who belonged to the Democratic Socialists of America (DSA) and had supported Bernie Sanders in the 2016 Democratic primaries, leaped into the national consciousness in June 2018 when she defeated ten-term incumbent Representative Joe Crowley, the chair of the House Democratic Caucus, in the Democratic primary; she went on to win the general election with 78 percent of the vote.[63] Although Ocasio-Cortez's victory came in a primary in a solidly Democratic district, it heralded (together with the similar primary victory of Ayanna Pressley over incumbent Michael Capuano in Boston) a strong public desire for new faces in government, a desire that helped lead to the large Democratic victory in the 2018 elections to the House of Representatives.

Ocasio-Cortez campaigned on a platform of Medicare for all, federal job creation, and abolition of the federal Immigration and Customs Enforcement (ICE) agency. However, she soon took up the idea of the Green New Deal and made it her own, working with movement groups such as 350.org and the Sunrise Movement to pressure other members of Congress to endorse it; many have done so, including several candidates for the Democratic presidential nomination.[64]

Support for the idea of a Green New Deal grew faster than the development of detailed policy proposals, leading to some skepticism among Democratic Party insiders. House Speaker Nancy Pelosi said in a February 6 interview, "It will be one of several or maybe many suggestions that we receive . . . the green dream or whatever they call it, nobody knows what it is, but they're for it right?" The next morning, Ocasio-Cortez and Senator Ed Markey (D-MA) introduced a congressional resolution providing a detailed list of what the Green New Deal consisted of. Pelosi became more supportive, responding to a question in her morning press conference, "Quite frankly I haven't seen it, but I do know that it's enthusiastic, and we welcome all the enthusiasms that are out there." She added:

> The Green New Deal points out that the public is much more aware of the challenge that we face, and that is a good thing, because the public sentiment will help us pass the most bold—common-denominator bold— initiatives, with an interest in, again, saving the planet while we create jobs, protect the health of our children, and pass the planet on in a very serious way.[65]

Representative Jim McGovern (D-MA), the chair of the House Rules Committee, is a cosponsor of the resolution and stood next to Ocasio-Cortez at the launch event. Since the Rules Committee clears bills and resolutions for votes on the House floor, the support of McGovern will be important.

The resolution cites the October 2018 special report of the IPCC on the need to keep global warming below two degrees Celsius to avoid dire results and relates this crisis to such others as declining life expectancy, four decades of economic stagnation and rising inequality, and large gaps in wealth and income based on race and gender. Citing the success of government mobilizations during the New Deal and the Second World War in creating "the greatest middle class that the United States has ever seen," the resolution calls for a twenty-year "Green New Deal mobilization" designed to create millions of good jobs, bring unprecedented prosperity, and remedy economic injustices. The resolution offers a detailed list of objectives, including eliminating fossil fuel–based electricity, weatherizing all buildings, building a sustainable food system based on family farming, eliminating fossil fuels from transportation, and providing everyone with high quality health care and housing.[66]

The current legislative form of Green New Deal is a resolution, not a bill; its passage would declare the sense of Congress that "it is the duty of the Federal Government to create a Green New Deal" along the lines spelled out in the last ten pages of the fourteen-page resolution. The intent is to provide a focus within Congress for action by grassroots movements. The 2018 Congressional election saw a high level of activism and participation within the Democratic Party, resulting not only in the gain of forty seats in the House but in exciting losses, particularly those of Stacey Abrams for governor of Georgia, Andrew Gillum for governor of Florida, and Beto O'Rourke for U.S. Senate from Texas. These close but losing campaigns made such a strong impression that Abrams was chosen to give the Democratic response to President Trump's 2019 State of the Union address, and O'Rourke is being considered seriously as a possible presidential candidate. The activists who drove these campaigns can now work to persuade their own representatives to support the resolution and recruit candidates to run against those who do not support it.

The campaign for a Green New Deal will be hard, and success is by no means assured. However, given the growing immediacy of the climate crisis, as detailed both by the IPCC and the United States' own National Climate Assessment on the one hand and the unwillingness of the congressional Republican Party to act on climate on the other, it is difficult to see an alternative path that would be adequate. It is always possible that at some point the dominant industrial and financial interests in the United States will see that their own future is threatened by climate change, and they will then wrest control of the Republican Party from the hands of the coal and oil industries. Until that happens, and maybe even if it does, grassroots politics appears to be the only possible path to planetary salvation.

Notes

Chapter 1

1. John Dodson, Xiaoqiang Li, Nan Sun, Pia Atahan, Zhou Xinying, Hanbin Liu, Keliang Zhao, Songmei Hu, and Zemeng Yang, "Use of Coal in the Bronze Age in China," *Holocene* 24, no. 5 (2014): 525–30.

2. Georgia L. Irby-Massie and Paul T. Keyser, *Greek Science of the Hellenistic Era: A Sourcebook* (London: Routledge, 2002), 228.

3. Dodson et al., "Use of Coal in the Bronze Age in China."

4. Richard Heinberg, *Blackout: Coal, Climate and the Last Energy Crisis* (Gabriola Island, BC: New Society, 2009), 3. For similar data for the United States, see Hobart King, "History of Energy Use in the United States," Geology.com, http://geology.com/articles/history-of-energy-use/.

5. Vaclav Smil, *Energy Transitions: History, Requirements, Prospects* (Santa Barbara: Praeger, 2010), 155. An exajoule is equal to 10^{18} joules, or a bit less than 950 trillion BTUs.

6. Dwight E. Collins, Russell M. Genet, and David Christian, "Crafting a New Narrative to Support Sustainability," in *State of the World 2013: Is Sustainability Still Possible?*, ed. Linda Starke (Washington, DC: Island Press, 2013), 220.

7. Collins, Genet, and Christian, "Crafting a New Narrative to Support Sustainability," 220.

8. George P. Marsh, *Man and Nature, or Physical Geography as Modified by Human Action* (New York: Charles Scribner, 1864). The quotation is taken from the excerpted passage in Daniel G. Payne and Richard S. Newman, eds., *The Palgrave Environmental Reader* (New York: Palgrave Macmillan, 2005), 80.

9. This argument is developed more fully in my earlier paper "Leave It in the Ground: Science, Politics, and the Movement to End Coal Use" (American Politics Group, University of Oxford, January 5–7, 2014), http://papers.ssrn.com/sol3/papers.cfm?abstract_id=2375370.

10. The Pinkerton Agency began as a private detective firm and now is primarily a provider of armored cars and security guards, but at one time it

operated its own private army, used mainly to break strikes. See James D. Horan, *The Pinkertons: The Detective Dynasty That Made History* (New York: Crown, 1968) and Frank Morn, *"The Eye That Never Sleeps": A History of the Pinkerton National Detective Agency* (Bloomington: Indiana University Press, 1982). For a strongly critical view of the agency, see Ward Churchill, "From the Pinkertons to the PATRIOT Act: The Trajectory of Political Policing in the United States, 1870 to the Present," *CR: The New Centennial Review* 4, no. 1 (2004): 1–72.

11. For detailed description and analysis of the Clean Air Act, see Charles O. Jones, *Clean Air: The Policies and Politics of Pollution Control* (Pittsburgh: University of Pittsburgh Press, 1975); for its subsequent development, see Gary C. Bryner, *Blue Skies, Green Politics: The Clean Air Act of 1990 and Its Implementation* (Washington, DC: CQ Press, 1995).

12. Actual greenhouses heat up much faster because their solid walls also block the heat from escaping by convection.

13. Bill McKibben, *The End of Nature* (New York: Random House, 1989). For Hansen's account of his testimony, see James Hansen, *Storms of My Grandchildren: The Truth About the Coming Climate Catastrophe and Our Last Chance to Save Humanity*, illustrations by Makiko Sato (New York: Bloomsbury USA, 2009).

Chapter 2

1. Émile Zola, *Germinal*, Les Rougon-Macquart (Paris: G. Charpentier, 1887).

2. Sean Patrick Adams, "The US Coal Industry in the Nineteenth Century," EH.net Encyclopedia, edited by Robert Whaples, January 23, 2003, http://eh.net/?s=coal+industry.

3. Adams, "The US Coal Industry in the Nineteenth Century."

4. Richard O. Boyer and Herbert M. Morais, *Labor's Untold Story.* (1955; repr., New York: United Electrical Workers, 1970), 44–47.

5. Boyer and Morais, *Labor's Untold Story*, 51; J. Anthony Lukas, *Big Trouble: A Murder in a Small Western Town Sets Off a Struggle for the Soul of America* (New York: Simon & Schuster, 1997), 179–84.

6. Adams, "The US Coal Industry in the Nineteenth Century."

7. Thomas G. Andrews, *Killing for Coal: America's Deadliest Labor War* (Cambridge, MA: Harvard University Press, 2008); George S. McGovern and Leonard F. Guttridge, *The Great Coalfield War*, maps by Samuel H. Bryant (Boston: Houghton Mifflin, 1972); Scott Martelle, *Blood Passion: The Ludlow Massacre and Class War in the American West* (New Brunswick, NJ: Rutgers University Press, 2007).

8. Robert H. Zieger, *The CIO: 1935–1955* (Chapel Hill: University of North Carolina Press, 1995). See also Robert H. Zieger, *John L. Lewis: Labor Leader*, Twentieth-Century American Biography Series, no. 8 (Woodbridge, CT: Twayne, 1988). See also Saul Alinsky, *John L. Lewis: An Unauthorized Biography* (New York: Putnam, 1949).

9. Nelson Lichtenstein, "Two Roads Forward for Labor: The AFL-CIO's New Agenda," *Dissent* 61, no. 1 (2014): 54–58; Chris Cillizza, "Trumka Hopes to Mend the AFL-CIO," *Washington Post*, July 13, 2009.

10. From a speech delivered January 12, 2012; quoted in Jeremy Brecher, "Stormy Weather: Climate Change and a Divided Labor Movement," *New Labor Forum* 22, no. 1 (2013): 76.

11. Nora Räthzel and David Uzzell, "Trade Unions and Climate Change: The Jobs versus Environment Dilemma," *Global Environmental Change* 21, no. 4 (October 2011): 1215–23. See also John C. Berg, ed., *Teamsters and Turtles? U.S. Progressive Political Movements in the 21st Century* (Lanham, MD: Rowman & Littlefield, 2003).

12. Chad Montrie, *To Save the Land and People: A History of Opposition to Surface Coal Mining in Appalachia* (Chapel Hill: University of North Carolina Press, 2003), 17.

13. Andrew Schissler, "Strip Mining," *Encyclopedia of Earth* (2006), http://www.eoearth.org/view/article/156280/.

14. Montrie, *To Save the Land and People*, 13–14; Wilma Dunaway, "Speculators and Settler Capitalists: Unthinking the Mythology about Appalachian Landholding, 1790–1860," in *Appalachia in the Making: The Mountain South in the Nineteenth Century*, eds. Mary Beth Pudup, Dwight B. Billings, and Altina L. Waller (Chapel Hill: University of North Carolina Press, 1995).

15. Montrie, *To Save the Land and People*, 15.

16. Harry M. Caudill, *Night Comes to the Cumberlands: A Biography of a Depressed Area* (Boston: Little, Brown, 1962), 72–75.

17. Montrie, *To Save the Land and People*, 185.

18. Marsh, *Man and Nature*; Gifford Pinchot, *The Fight for Conservation* (London: Hodder & Stoughton, 1910).

19. John Muir, *A Brief Statement of the Hetch-Hetchy Case to Date* (n.p., 1909),http://lcweb2.loc.gov/gc/amrvg/vg50/vg50.html; John Muir, *Meditations of John Muir: Nature's Temple*, compiled and edited by Chris Highland (Berkeley: Wilderness Press, 2001). See also Robert Marshall, "The Wilderness as a Minority Right," *U.S. Forest Service Bulletin*, August 27, 1928, 5–6; however, this isolated quotation oversimplifies Marshall's views. Gottlieb argues persuasively that Marshall's aim was a "democratic wilderness," in which public ownership of the forests would assure that they would be used for social welfare. See Robert Gottlieb, *Forcing the Spring: The Transformation of the American Environmental Movement,* rev. ed. (Washington, DC: Island Press, 2005), 47–51.

20. Montrie, *To Save the Land and People*, 202.

21. Judi Bari, *Timber Wars* (Monroe, ME: Common Courage, 1994) provides some positive instances of logger opposition to clear-cutting forests.

22. Montrie, *To Save the Land and People*, 178.

23. Montrie, *To Save the Land and People*, 189–90.

24. Jim Schwab, *Deeper Shades of Green: The Rise of Blue-Collar and Minority Environmentalism in America* (San Francisco: Sierra Club Books, 1994), 290–91.

25. Montrie, *To Save the Land and People*, 196; Schwab, *Deeper Shades of Green*, 298–99.

26. Montrie, *To Save the Land and People*, 199.

27. Andrews, *Killing for Coal*.

28. Berg, "Leave It in the Ground," paper.

29. "Classical environmentalism" has been used to describe a political process in which scientists present evidence of an environmental problem, national organizations rally public support to demand that something be done about the problem, and Congress and the president act to resolve the problem. See Kai N. Lee, William R. Freudenburg, and Richard B. Howarth, *Humans in the Landscape: An Introduction to Environmental Studies* (New York: W. W. Norton, 2013).

Chapter 3

1. David Stradling and Peter Thorsheim, "The Smoke of Great Cities: British and American Efforts to Control Air Pollution, 1860–1914," *Environmental History* 4, no. 1 (January 1999): 8.

2. Peter Brimblecombe, *The Big Smoke: A History of Air Pollution in London since Medieval Times* (London: Methuen, 1987), 92–95.

3. Brimblecombe, *Big Smoke*, 153–55, 175–76.

4. Devra Davis, *When Smoke Ran Like Water: Tales of Environmental Deception and the Battle against Pollution* (Reading, MA: Basic Books, 2002), 1–13.

5. Davis, *When Smoke Ran Like Water*, 27–28.

6. Davis, *When Smoke Ran Like Water*, 15–19.

7. Quoted in Davis, *When Smoke Ran Like Water*, 18.

8. Jeanne R. Lowe, *Cities in a Race with Time: Progress and Poverty in America's Renewing Cities* (New York: Random House, 1967), 138, quoted in Jones, *Clean Air*, 46.

9. Brimblecombe, *Big Smoke*, 9.

10. Brimblecombe, *Big Smoke*, 66.

11. Brimblecombe, *Big Smoke*, 30,

12. Brimblecombe, *Big Smoke*, 7.

13. Davis, *When Smoke Ran Like Water*, 41–42.

14. John Evelyn, *Fumifugium: Or, The Inconvenience of the Aer and Smoak of London Dissipated* (Manchester: National Smoke Abatement Society, 1933), cited in Davis, *When Smoke Ran Like Water*, 35–36.

15. E. T. Wilkins, "Air Pollution and the London Fog of December, 1952," *Journal of the Royal Sanitary Institute* 74, no. 1 (January 1954): 3–6.

16. Wilkins, "Air Pollution and the London Fog of December, 1952," 10.

17. Wilkins, "Air Pollution and the London Fog of December, 1952," 11–13. Contemporary public health reports attributed this second wave of deaths to a flu epidemic, rather than the fog; however, subsequent analysis has shown that the air pollution was a major factor. See Michelle L. Bell and Devra Lee Davis, "Reassessment of the Lethal London Fog of 1952: Novel Indicators of Acute and Chronic Consequences of Acute Exposure to Air Pollution," *Environmental Health Perspectives* 109, no. S3 (June 2001): S389–94.

18. Scott Hamilton Dewey, *Don't Breathe the Air: Air Pollution and U.S. Environmental Politics, 1945–1970*, Environmental History Series, no. 16. (College Station: Texas A&M University Press, 2000), 37–38.

19. Dewey, *Don't Breathe the Air*, 39.

20. Dewey, *Don't Breathe the Air*, 38.

21. Dewey, *Don't Breathe the Air*, 38–39.

22. Benjamin Ross and Steven Amter, *The Polluters: The Making of Our Chemically Altered Environment* (Oxford: Oxford University Press, 2010), 73.

23. The story is told in Ross and Amter, *The Polluters*, 79–80.

24. Raymond R. Tucker, *The Los Angeles Smog Report* (Los Angeles: Times-Mirror, 1947) See also Dewey, *Don't Breathe the Air*, 42–43.

25. Ross and Amter, *The Polluters*, 78–79.

26. Dewey, *Don't Breathe the Air*, 42–44; Ross and Amter, *The Polluters*, 79–81.

27. Both quoted in Dewey, *Don't Breathe the Air*, 47; see also James E. Krier and Edmund Ursin, *Pollution and Policy: A Case Essay on California and Federal Experience with Motor Vehicle Air Pollution, 1940–1975* (Berkeley: University Of California Press, 1977), 59, 73–75; Marvin Brienes, "The Fight against Smog in Los Angeles, 1943–1957" (PhD diss., University of California Davis, 1975), 124, 161–64, 192.

28. Ross and Amter, *The Polluters*, 83–84; Arie Haagen-Smit, "Formation of Ozone in Los Angeles Smog," in *Proceedings of the Second National Air Pollution Symposium* (Pasadena, May 5–6, 1952), 54–56.

29. Louis McCabe, ed., *Air Pollution: Proceedings of the United States Technical Conference on Air Pollution*, sponsored by the Interdepartmental Committee on Air Pollution (New York: McGraw-Hill, 1952), v-vii, quoted in Jones, *Clean Air*, 30.

30. Public Law 159, July 14, 1955, quoted in Jones, *Clean Air*, 31.

31. See, for example, John Locke, "The Second Treatise," in *Two Treatises of Government*, ed. Peter Laslett, reprint, 1960 (New York: Mentor Books, 1965).

32. For a full account of this philosophical evolution see David J. Bodenhamer, *The Revolutionary Constitution* (New York: Oxford University Press, 2012).

33. See Paul R. Benson, *The Supreme Court and the Commerce Clause, 1937–1970*, foreword by Maurice G. Baxter (New York: Dunellen, 1970).

34. Jones, *Clean Air*, 33–36.

35. Jones, *Clean Air*, 62–63.

36. Randall B. Ripley, "Congress and Clean Air: The Issue of Enforcement, 1963," in *Congress and Urban Problems: A Casebook on the Legislative Process*, ed. Frederic N. Cleaveland (Washington, DC: Brookings Institution, 1969), 237.

37. Ripley, "Congress and Clean Air," 238–49; quoted passage at 248.

38. Ripley, "Congress and Clean Air," 251.

39. Ripley, "Congress and Clean Air," 255–58; Roberts quoted at 255.

40. Ripley, "Congress and Clean Air," 250.

41. Ripley, "Congress and Clean Air," 261–74.

42. Jones, *Clean Air*, 74–75, 120.

43. Johnson's 1967 State of the Union address, quoted in Jones, *Clean Air*, 79.

44. Ozone in the stratosphere has the beneficial quality of blocking dangerous ultraviolet radiation. At ground level, it can damage the eyes and the

respiratory tract in humans and other animals; it is one of the major components of smog. The ozone isotope is too unstable to survive the thirty thousand–foot climb from ground to stratosphere, so the two ozone populations are not related to each other. Bryner, *Blue Skies, Green Politics*, 43.

45. Walter A. Rosenbaum, *Environmental Politics and Policy,* 10th ed. (Washington, DC: CQ Press, 2017), 201.

46. For a full account of the controversy, see Jones, *Clean Air,* 253–72.

47. Gene E. Likens, F. Herbert Bormann, and Noye M. Johnson, "Acid Rain," *Environment: Science and Policy for Sustainable Development* 14, no. 2 (March 1, 1972): 33–40; Gene E. Likens and F. Herbert Bormann, "Acid Rain: A Serious Regional Environmental Problem," *Science* 184, no. 4142 (June 14, 1974): 1176–79.

48. Bruce A. Ackerman and William T. Hassler, *Clean Coal/Dirty Air: Or, How the Clean Air Act Became a Multibillion-Dollar Bail-Out for High-Sulfur Coal Producers and What Should Be Done About It* (New Haven: Yale University Press, 1981).

49. There are many examples of this conflict in the literature; see, for example, Dewey, *Don't Breathe the Air*, 76–77 and Bryner, *Blue Skies, Green Politics*, 88–92.

50. Jones, *Clean Air,* 238.

51. Henry A. Waxman, "An Overview of the Clean Air Act Amendments of 1990," *Environmental Law* 21, no. 4 (1991), 1742.

52. Bryner, *Blue Skies, Green Politics*, 126.

53. Bryner, *Blue Skies, Green Politics*, 126–27.

54. See Thomas E. Mann and Norman J. Ornstein, *It's Even Worse Than It Looks: How the American Constitutional System Collided with the New Politics of Extremism* (New York: Basic Books, 2013) for a detailed account of the growing partisanship of American politics in general, and Congress in particular.

Chapter 4

1. Water vapor also acts as a greenhouse gas. However, most water enters the atmosphere through evaporation from the ocean, rather than by emissions from other sources. The amount of atmospheric water is determined by the pressure and temperature at any one point. Therefore, it is not included among those greenhouse gases that may be controlled.

2. Pamela S. Chasek, David L. Downie, and Janet Welsh Brown, *Global Environmental Politics,* 7th ed., Dilemmas in World Politics (Boulder: Westview Press, 2017), 267.

3. Nils Ekholm, "On the Variations of the Climate of the Geological and Historical Past and Their Causes," *Quarterly Journal of the Royal Meteorological Society* 27, no. 117 (1901): 1–62; J. H. Poynting, "On Prof. Lowell's Method for Evaluating the Surface-Temperatures of the Planets; with an Attempt to Represent the Effect of Day and Night on the Temperature of the Earth," *Philosophical Magazine* 14, no. 84 (December 1907): 749–60. A blog post by Steve Easterbrook puts together the story of the development of the term: Steve Easterbrook, "Who

First Coined the Term 'Greenhouse Effect'?," *Serendipity* (blog), August 18, 2015, http://www.easterbrook.ca/steve/2015/08/who-first-coined-the-term-greenhouse-effect/.

4. Revelle had previously published some theoretical calculations but had found the data insufficient to reach a conclusion. See Roger Revelle and Hans E. Suess, "Carbon Dioxide Exchange between Atmosphere and Ocean and the Question of an Increase of Atmospheric CO_2 during the Past Decades," *Tellus* 9, no. 1 (1957): 18–27.

5. Charles D. Keeling, "The Concentration and Isotopic Abundances of Carbon Dioxide in the Atmosphere," *Tellus* 12, no. 2 (1960): 203.

6. Lee, Freudenburg, and Howarth, *Humans in the Landscape.*

7. Hansen, *Storms of My Grandchildren*, xv.

8. Philip Shapecoff, "Global Warming Has Begun, Expert Tells Senate," *New York Times*, June 24, 1988, http://www.nytimes.com/1988/06/24/us/global-warming-has-begun-expert-tells-senate.html.

9. Rosenbaum, *Environmental Politics and Policy,* 10th ed., 366–67.

10. Details can be found at the IPCC website, http://www.ipcc.ch. See also Rosenbaum, *Environmental Politics and Policy,* 10th ed., 359–60.

11. Intergovernmental Panel on Climate Change, "Policymaker Summary of Working Group I (Scientific Assessment of Climate Change)," in *Climate Change: The IPCC 1990 and 1992 Assessments, First Assessment Report Overview and Policymaker Summaries and 1992 IPCC Supplement* (Geneva: World Meteorological Association and United Nations Environment Programme, 1992), https://www.ipcc.ch/report/climate-change-the-ipcc-1990-and-1992-assessments/.

12. Lydia Saad, "Global Warming Concern at Three-Decade High in U.S.," *Gallup*, March 14, 2017, http://www.gallup.com/poll/206030/global-warming-concern-three-decade-high.aspx.

13. Tom Boden, Bob Andres, and Gregg Marland, *Global CO_2 Emissions from Fossil-Fuel Burning, Cement Manufacture, and Gas Flaring: 1751–2014* (Washington, DC: Oak Ridge National Laboratory, 2017), http://cdiac.ornl.gov/ftp/ndp030/global.1751_2014.ems.

14. Boden, Andres, and Marland, *Global CO_2 Emissions.*

15. Intergovernmental Panel on Climate Change, "Observations: Atmosphere and Surface," in *Climate Change 2013: The Physical Science Basis. Contribution of Working Group I to the Fifth Assessment Report of the Intergovernmental Panel on Climate Change* (Cambridge, UK: Cambridge University Press, 2013), 165.

16. Environmental Protection Agency, "Climate Change Indicators: Atmospheric Concentrations of Greenhouse Gases," updated April 2016, https://www.epa.gov/climate-indicators/climate-change-indicators-atmospheric-concentrations-greenhouse-gases.

17. Ed Dlugokencky and Pieter Tans, "Trends in Atmospheric Carbon Dioxide: Recent Global CO_2," Earth System Research Laboratory, National Oceanic and Atmospheric Administration, viewed July 25, 2017, https://www.esrl.noaa.gov/gmd/ccgg/trends/gl_data.html.

18. Rebecca Lindsey and LuAnn Dahlman, "Climate Change: Global Temperature," Climate.gov, updated August 1, 2018, https://www.climate.gov/news-features/understanding-climate/climate-change-global-temperature.

19. NOAA National Centers for Environmental information, "Climate at a Glance: Global Time Series," viewed July 25, 2017, http://www.ncdc.noaa.gov/cag/global/time-series.

20. For example, see James Hansen et al., "Global Surface Temperature Change," *Reviews of Geophysics* 48, no. 4 (2010), https://agupubs.onlinelibrary.wiley.com/doi/epdf/10.1029/2010RG000345.

21. P. J. Webster et al., "Changes in Tropical Cyclone Number, Duration, and Intensity in a Warming Environment," *Science* 309, no. 5742 (September 15, 2005): 1844.

22. C. D. Hoyos et al., "Deconvolution of the Factors Contributing to the Increase in Global Hurricane Intensity," *Science* 312, no. 5770 (April 7, 2006): 94.

23. Webster et al., "Changes in Tropical Cyclone Number, Duration, and Intensity."

24. Megan Suzanne Mallard, "Atlantic Hurricanes and Climate Change" (PhD diss., North Carolina State University, 2011), 4–5, http://www.lib.ncsu.edu/resolver/1840.16/7184. While the upper troposphere is warmed more than the lower troposphere, the stratosphere—the next layer up—is actually cooled.

25. The 30 percent estimate is from Richard A. Feely et al., "Impact of Anthropogenic CO_2 on the $CaCO_2$ System in the Oceans," *Science* 305, no. 5682 (July 15, 2004): 362; the 25 percent estimate is from PMEL Carbon Program, National Oceanic and Atmospheric Administration, "Ocean Acidification: The Other Carbon Dioxide Problem," https://pmel.noaa.gov/co2/story/Ocean+Acidification.

26. PMEL Carbon Program. National Oceanic and Atmospheric Administration, "What is Ocean Acidification?," https://pmel.noaa.gov/co2/story/What+is+Ocean+Acidification%3F.

27. Feely et al., "Impact of Anthropogenic CO_2 on the $CaCO_2$ System in the Oceans," 362.

28. Ken Caldeira and Michael E. Wickett, "Oceanography: Anthropogenic Carbon and Ocean pH," *Nature* 425, no. 6956 (September 25, 2003): 365.

29. Clayton M. Christensen, *The Innovator's Dilemma: When New Technologies Cause Great Firms to Fail* (New York: Harper Business, 2000). See also Clayton M. Christensen and Michael E. Raynor, *The Innovators Solution: Creating and Sustaining Successful Growth* (Boston: Harvard Business School, 2003).

30. U.S. Energy Information Administration, *U.S. Coal Reserves* (2016), https://www.eia.gov/coal/reserves/.

31. U.S. Energy Information Administration, *Annual Coal Report 2015* (November 2016), https://www.eia.gov/coal/annual/.

32. Center for Responsive Politics, "Coal Mining: Long-Term Contribution Trends," OpenSecrets.org, http://www.opensecrets.org/industries/totals.php?cycle=2018&ind=E1210.

33. Kevin Grandia, "Leaked Clean Coal Strategy Memo to Peabody Energy," *DeSmog* (blog), January 16, 2009, https://www.desmogblog.com/leaked-clean -coal-strategy-memo-peabody-energy.

34. Stephen L. Miller, "Coal Industry Strategy Letter to CEO of Peabody Energy," June 18, 2004, https://www.desmogblog.com/sites/beta.desmogblog. com/files/Coal%20Industry%20Strategy%20Letter%20To%20CEO%20of%20 Peabody%20Energy.pdf.

35. Brad Jones, "New Multi-Industry Coalition Aligns to Advocate Energy Security and Environmental Stewardship," Press release, American Coalition for Clean Coal Electricity, April 17, 2008, https://www.businesswire.com/news /home/20080417006084/en/New-Multi-Industry-Coalition-Aligns-Advocate -Energy-Security.

36. Marianne Lavelle, "The 'Clean Coal' Lobbying Blitz: As Climate Change Hearings Begin on Capitol Hill, a Coal Industry Group Flexes New-Found Muscle," *Center for Public Integrity*, April 21, 2009, https://www.publicintegrity. org/2009/04/21/2885/clean-coal-lobbying-blitz. See also Anne C. Mulkern, "'Citizen Army' Carries Coal's Climate Message to Hinterlands," *New York Times*, August 6, 2009, http://www.nytimes.com/gwire/2009/08/06/06greenwire-citizen -army-carries-coals-climate-message-to-39075.html.

37. Brian McNeill, "Forged Letters Scandal Widens," *Daily Progress*, August 5, 2009, https://www.dailyprogress.com/news/forged-letters-scandal-widens/article _7dfdb333-4860-57b2-89fd-4852ac6425fe.html.

38. National Council for Science and the Environment, "Building Climate Solutions," 14th National Conference and Global Forum on Science, Policy and the Environment, January 28–31, 2014, Washington, DC. The conference program that includes Gummer and Inglis can be found online at https://ncsecon ference.org/wp-content/uploads/2017/06/2014-conference-program.pdf.

39. Thomas A. Schatz, public comment on "Regulating Greenhouse Gases under the Clean Air Act," No. EPA-HQ-OAR-2008–0318, n.d., 1–2, https:// www.heartland.org/_template-assets/documents/publications/24254.pdf.

40. Schatz, "Regulating Greenhouse Gases under the Clean Air Act," 2.

41. Mario Lewis et al., public comment on "Regulating Greenhouse Gases under the Clean Air Act," No. EPA-HQ-OAR-2008–0318, November 24, 2008, 1, https://www.heartland.org/_template-assets/documents/publications/24270.pdf.

42. Although the statement is often quoted, I have not found a direct source. This version comes from a column by James K. Glassman, "Administration in the Balance," *Wall Street Journal*, Eastern edition, March 8, 2001.

43. The average global temperature has risen more rapidly than predicted over the last twenty years. See Lindsey and Dahlman, "Climate Change: Global Temperature."

44. J. T. Houghton et al., eds., *Climate Change 2001: The Scientific Basis, Contribution of Working Group I to the Third Assessment Report of the Intergovernmental Panel on Climate Change* (Cambridge, UK: Cambridge University Press, 2001).

45. Suzanne Goldenberg and Helena Bengtsson, "Biggest US Coal Company Funded Dozens of Groups Questioning Climate Change," *Guardian*, June 13, 2016, https://www.theguardian.com/environment/2016/jun/13/peabody-energy-coal-mining-climate-change-denial-funding.

46. Richard S. Lindzen, "Some Coolness Concerning Global Warming," *Bulletin of the American Meteorological Society* 71, no. 3 (March 1, 1990): 290.

47. Richard S. Lindzen, "Climate Physics, Feedbacks, and Reductionism (and When Does Reductionism Go Too Far?)," *European Physical Journal Plus* 127, no. 5 (2012): Article 52.

48. The Hadley circulation is a pattern whereby hot air rises from the equator for ten to fifteen kilometers and then moves toward the two poles and sinks to the surface again in the subtropics.

49. Richard S. Lindzen, "Can Increasing Carbon Dioxide Cause Climate Change?," *PNAS: Proceedings of the National Academy of Sciences of the United States of America* 94, no. 16 (August 5, 1997): 8335, http://www.pnas.org/content/94/16/8335.

50. Richard S. Lindzen, Ming-Dah Chou, and Arthur Y. Hou, "Does the Earth Have an Adaptive Infrared Iris?," *Bulletin of the American Meteorological Society* 82, no. 3 (March 2001): 417.

51. Dennis L. Hartmann and Marc L. Michelsen, "No Evidence for Iris," *Bulletin of the American Meteorological Society* 83, no. 2 (February 2002): 249.

52. Richard S. Lindzen, Ming-Dah Chou, and Arthur Y. Hou, "Comment on 'No Evidence for Iris,'" *Bulletin of the American Meteorological Society* 83, no. 9 (September 2002): 1345–49.

53. Bing Lin et al., "The Iris Hypothesis: A Negative or Positive Cloud Feedback?," *Journal of Climate* 15, no. 1 (January 2002): 3.

54. Ming-Dah Chou, Richard S. Lindzen, and Arthur Y. Hou, "Comments on 'The Iris Hypothesis: A Negative or Positive Cloud Feedback?,'" *Journal of Climate* 15, no. 18 (September 15, 2002): 2713–15.

Chapter 5

1. President Donald Trump is a possible exception. Trump said during the 2016 campaign that climate change was a hoax; at other times, however, he has acknowledged that it is a problem. The Trump administration's position will be discussed in a later chapter.

2. Converted from the figure of 3,755 million tonnes of oil equivalent (Mtoe) in International Energy Agency, *World Energy Outlook 2017* (Paris: International Energy Agency, 2017), 79, http://www.iea.org/weo2017/. One petawatt is equal to 1 million megawatts or 1 billion kilowatts.

3. Bryner, *Blue Skies, Green Politics*.

4. Miller, "Coal Industry Strategy Letter."

5. Barack Obama, *Presidential Memorandum: A Comprehensive Federal Strategy on Carbon Capture and Storage* (Washington, DC: White House, Office of

the Press Secretary, 2010),https://obamawhitehouse.archives.gov/realitycheck/the-press-office/presidential-memorandum-a-comprehensive-federal-strategy-carbon-capture-and-storage.

6. David Biello, "The Carbon Capture Fallacy," *Scientific American* 314, no. 1 (January 2016): 58–65.

7. Gary T. Rochelle, "Amine Scrubbing for CO_2 Capture," *Science* 325, no. 5948 (September 25, 2009): 1652.

8. Rochelle, "Amine Scrubbing for CO_2 Capture," 1653.

9. The scientifically minded will recognize the first and second laws of thermodynamics here.

10. Henry Fountain, "Turning Carbon Dioxide Into Rock, and Burying It," *New York Times*, February 9, 2015, https://www.nytimes.com/2015/02/10/science/burying-a-mountain-of-co2.html.

11. Henry Fountain, "Corralling Carbon before It Belches from Stack," *New York Times*, July 21, 2014, https://www.nytimes.com/2014/07/22/science/corralling-carbon-before-it-belches-from-stack.html.

12. Ian Austen, "Technology to Make Clean Energy from Coal Is Stumbling in Practice," *New York Times*, March 29, 2016, https://www.nytimes.com/2016/03/30/business/energy-environment/technology-to-make-clean-energy-from-coal-is-stumbling-in-practice.html.

13. David Roberts, "Turns Out the World's First 'Clean Coal' Plant Is a Backdoor Subsidy to Oil Producers," *Grist*, March 31, 2015, https://grist.org/climate-energy/turns-out-the-worlds-first-clean-coal-plant-is-a-backdoor-subsidy-to-oil-producers/.

14. Global CCS Institute, *The Global Status of CCS: 2017* (Melbourne: Global CCS Institute, 2017), 52.

15. Global CCS Institute, *Projects Database: Large-Scale CCS Facilities* (Melbourne: Global CCS Institute, 2018), http://www.globalccsinstitute.com/projects/large-scale-ccs-projects.

16. Megan Geuss, "$7.5 Billion Kemper Power Plant Suspends Coal Gasification," *Ars Technica*, June 28, 2017, https://arstechnica.com/information-technology/2017/06/7-5-billion-kemper-power-plant-suspends-coal-gasification/; James Conca, "The Largest Clean Coal Power Plant in America Turns to Natural Gas," *Forbes*, July 11, 2017, https://www.forbes.com/sites/jamesconca/2017/07/11/the-largest-clean-coal-power-plant-in-america-turns-to-natural-gas/#26c8916f7d26.

17. Ian Urbina, "Piles of Dirty Secrets behind a Model 'Clean Coal' Project," *New York Times*, July 5 2016, https://www.nytimes.com/2016/07/05/science/kemper-coal-mississippi.html.

18. Sharon Kelly, "Bosses at World's Most Ambitious Clean Coal Plant Kept Problems Secret for Years," *Guardian*, March 2, 2018, https://www.theguardian.com/us-news/2018/mar/02/clean-coal-kemper-plant-mississippi-problems; Sharon Kelly, "How America's Clean Coal Dream Unravelled," *Guardian*, March 2, 2018, https://www.theguardian.com/environment/2018/mar/02/clean-coal-america-kemper-power-plant.

19. R. Stuart Haszeldine, "Carbon Capture and Storage: How Green Can Black Be?," *Science* 325, no. 5948 (September 25, 2009): 1648–49.

20. Haszeldine, "Carbon Capture and Storage: How Green Can Black Be?," 1651.

21. International Energy Agency, *Carbon Capture and Storage Roadmap* (n.d.), https://www.iea.org/publications/freepublications/publication/CCS_roadmap _foldout.pdf.

22. R. S. Lampitt et al., "Ocean Fertilization: A Potential Means of Geoengineering?," *Philosophical Transactions of the Royal Society A: Mathematical, Physical and Engineering Sciences* 366, no. 1882 (2008): 3919–45.

23. William C. G. Burns, "Geoengineering the Climate: An Overview of Solar Radiation Management Options," *Tulsa Law Review* 46, no. 2 (Winter 2010): 294–96.

24. Some people believe that this is happening now but kept secret by governments. The releases are referred to as "chemtrails." However, there is no credible evidence for this belief. See Daniel Loxton, "Terrifying! Improbable! Chemtrails!," *Junior Skeptic* 63, supplement to *Skeptic* 22, no. 2 (Spring 2017): S64; Rose Cairns, "Climates of Suspicion: 'Chemtrail' Conspiracy Narratives and the International Politics of Geoengineering," *Geographical Journal* 182, no. 1 (March 2016): 70–84; Alexandra Bakalaki, "Chemtrails, Crisis, and Loss in an Interconnected World," *Visual Anthropology Review* 32, no. 1 (May 2016): 12–23.

25. John Latham et al., "Global Temperature Stabilization via Controlled Albedo Enhancement of Low-Level Maritime Clouds," *Philosophical Transactions of the Royal Society A: Mathematical, Physical and Engineering Sciences* 366, no. 1882 (2008): 3969–87.

26. Alan Robock, Luke Oman, and Georgiy L. Stenchikov, "Regional Climate Responses to Geoengineering with Tropical and Arctic SO_2 Injections," *Journal of Geophysical Research: Atmospheres* 113, no. D16 (August 2008): 13.

27. Ulrike Niemeier and Simone Tilmes, "Sulfur Injections for a Cooler Planet," *Science* 357, no. 6348 (July 21, 2017): 246–48.

28. Vaclav Smil, *Energies: An Illustrated Guide to the Biosphere and Civilization* (Cambridge, MA and London: MIT Press, 1999), 6. An exajoule is one million petajoules.

29. Oliver Morton, "A New Day Dawning?: Silicon Valley Sunrise," *Nature* 443 (September 6, 2006): 19.

30. Calculated from figures in thousands of Megawatts from U.S. Energy Information Administration, "Electric Power Monthly," https://www.eia.gov /electricity/monthly/epm_table_grapher.php?t=epmt_1_1_a.

31. U.S. Energy Information Administration, "Electric Power Monthly."

32. International Energy Agency, *World Energy Outlook 2017*, 299.

33. Balthasar Henry Meyer and Caroline Elizabeth MacGill, et al., *History of Transportation in the United States before 1860*, prepared under the direction of Balthasar Henry Meyer by Caroline E. MacGill and a staff of collaborators, Contributions to American Economic History [3] (Washington, DC: Carnegie Institution of Washington, 1948).

34. Robert E. Pike, *Tall Trees, Tough Men* (New York: W. W. Norton, 1999); Glenn Richmond, "The Last Logjam: Machias River Log Drive," *Down East: The Magazine of Maine*, May 1971, https://downeast.com/archives-may-1971/.

35. Terry S. Reynolds, *Stronger than a Hundred Men: A History of the Vertical Water Wheel*, Johns Hopkins Studies in the History of Technology, no. 7 (Baltimore: Johns Hopkins University Press, 1983).

36. Steven Kreis, "Lecture 17: The Origins of the Industrial Revolution in England," *The History Guide*, last updated April 24, 2017, http://www.history guide.org/intellect/lecture17a.html; Patrick M. Malone, *Waterpower in Lowell: Engineering and Industry in Nineteenth-Century America*, Johns Hopkins Introductory Studies in the History of Technology (Baltimore: Johns Hopkins University Press, 2009).

37. REN21, *Renewables 2016: Global Status Report* (Paris: REN Secretariat, 2016), 32, http://www.ren21.net/wp-content/uploads/2016/06/GSR_2016_Full _Report_REN21.pdf.

38. International Energy Agency, *Renewable Energy Essentials: Hydropower*, 2010, http://www.iea.org/publications/freepublications/publication/hydropower _essentials.pdf.

39. International Energy Agency, *World Energy Outlook 2017*, 257.

40. World Commission on Dams, *Dams and Development: A New Framework for Decision-Making* (London: Earthscan, 2000), 104.

41. World Commission on Dams, *Dams and Development*, 259–75.

42. Kaifeng Li et al., "Problems Caused by the Three Gorges Dam Construction in the Yangtze River Basin: A Review," *Environmental Reviews* 21, no. 3 (September 2013): 127.

43. Itaipu Binacional, "Energy," https://www.itaipu.gov.br/en/energy/energy.

44. Trevor J. Price, "James Blyth: Britain's First Modern Wind Power Pioneer," *Wind Engineering* 29, no. 3 (May 1, 2005): 191–200.

45. Lauha Fried et al., *Global Wind Energy Outlook 2016* (Brussels: Global Wind Energy Council, 2016), 11.

46. Lauha Fried, ed., *Global Wind Statistics 2017* (Brussels: Global Wind Energy Council, 2018), 2.

47. For one example, see John C. Berg, "Cape Wind: A Case Study in the Politics of Technology Choice," in *Beyond the Global Village: Environmental Challenges Inspiring Global Citizenship*, eds. Rafaela C. Hillerbrand and Rasmus Karlsson (Oxford: Inter-Disciplinary Press, 2007), 65–74.

48. John Evelyn, *Sylva: or, A Discourse of Forest Trees and the Propagation of Timber* (London: J. Martyn, and J. Allestry, printers to the Royal Society, 1670).

49. Gabriel E. Lade, C-Y Cynthia Lin Lawell, and Aaron Smith, "Designing Climate Policy: Lessons from the Renewable Fuel Standard and the Blend Wall," *American Journal of Agricultural Economics* 100, no. 2 (March 2018): 587–88.

50. Lade, Lawell, and Smith, "Designing Climate Policy," 588.

51. U. S. Department of Energy. Alternative Fuels Data Center, "Renewable Fuel Standard," https://www.afdc.energy.gov/laws/RFS.

52. How this works is far beyond the scope of this book.

53. For classic statements of the argument, see Amory B. Lovins and John H. Price, *Non-Nuclear Futures: The Case for an Ethical Energy Strategy*, Friends of the Earth Energy Papers. (San Francisco: Friends of the Earth International, 1975); Anna Gyorgy and friends, *No Nukes: Everyone's Guide to Nuclear Power* (Boston: South End Press, 1979).

54. See James Lovelock, "Nuclear Power Is the Only Green Solution," *Independent*, May 23, 2004, https://www.independent.co.uk/voices/commentators /james-lovelock-nuclear-power-is-the-only-green-solution-564446.html; George Monbiot, "Why Fukushima Made Me Stop Worrying and Love Nuclear Power," *Guardian*, March 21, 2011, https://www.theguardian.com/commentisfree/2011 /mar/21/pro-nuclear-japan-fukushima.

55. Diane Cardwell, "The Murky Future of Nuclear Power in the United States," *New York Times*, February 18, 2017, https://www.nytimes.com/2017/02/18 /business/energy-environment/nuclear-power-westinghouse-toshiba.html.

56. International Energy Agency, *World Energy Outlook 2017*, 285–86.

57. William Stanley Jevons, *The Coal Question: An Enquiry Concerning the Progress of the Nation, and the Probable Exhaustion of Our Coal-Mines* (London: Macmillan, 1866).

58. Blake Alcott, "Jevons' Paradox," *Ecological Economics* 54, no. 1 (July 1, 2005): 9–21.

Chapter 6

1. Hardin's eponymous example concerned a group of farmers who each have a right to graze cattle on a common pasture. If each farmer seeks to maximize his or her self-interest, each will be driven to put so many cattle on the pasture that they will destroy it by overgrazing. See Garrett Hardin, "The Tragedy of the Commons," *Science* 162 (1968): 1243–48. The case is made more rigorously and generally by Mancur Olson Jr., *The Logic of Collective Action: Public Goods and the Theory of Groups*, 2nd ed. (Cambridge, MA: Harvard University Press, 1971).

2. See especially Elinor Ostrom, *Governing the Commons: The Evolution of Institutions for Collective Action* (Cambridge, UK: Cambridge University Press, 1990).

3. Quoted in Clyde Sanger, "Environment and Development," *International Journal* 28, no. 1 (1972): 105.

4. For details, see Sanger, "Environment and Development."

5. Robin Clarke and Lloyd Timberlake, *Stockholm Plus Ten: Promises, Promises? The Decade since the 1972 UN Environment Conference*, Earthscan Paperback (London: International Institute for Environment and Development, 1982), summarized in Gill Seyfang, "Environmental Mega-Conferences: From Stockholm to Johannesburg and Beyond," *Global Environmental Change* 13, no. 3 (2003): 224.

6. World Commission on Environment and Development, *Our Common Future* (Oxford: Oxford University Press, 1987).

7. Additional important documents included two other binding agreements, the Convention on Biological Diversity and the United Nations Convention to Combat Desertification. In addition, there were three major nonbinding statements of common purpose, the Rio Declaration on Environment and Development, Agenda 21, and the Authoritative Statement of Principles for a Global Consensus on the Management, Conservation, and Sustainable Development of All Types of Forests. The Commission on Sustainable Development was established to monitor the implementation of all these documents.

8. Union of International Associations, "Intergovernmental Negotiating Committee for a Framework Convention on Climate Change (INC/FCCC)," Open Yearbook, https://uia.org/s/or/en/1100025172.

9. Chasek, Downie, and Brown, *Global Environmental Politics*, 7th ed., 165–67. William Reilly, who was EPA administrator at the time, reports that he told Klaus Töpfer, the German Environment Minister, that a climate agreement would be tough politically for Bush and suggested a Forest Convention. Töpfer replied that Kohl was "nuts about trees" and would probably take the deal. See William K. Reilly, "Reflections on U.S. Environmental Policy: An Interview with William K. Reilly," 6, http://www.epaalumni.org/userdata/pdf/3E0FC143699 CD31B.pdf.

10. Edward A. Parson, "Assessing UNCED and the State of Sustainable Development," *Proceedings of the American Society of International Law Annual Meeting* 87 (1993): 509; Mark Valentine, "The Twelve Days of UNCED," *Earth Island Journal* 7, no. 3 (Summer 1992): 38–40; Irving M. Mintzer and J. Amber Leonard, eds., *Negotiating Climate Change: The Inside Story of the Rio Convention*, Cambridge Studies in Energy and Environment (Cambridge, UK: Cambridge University Press, 1994); Chasek, Downie, and Brown, *Global Environmental Politics*, 7th ed., 167.

11. Mintzer and Leonard, *Negotiating Climate Change*.

12. United Nations Framework Convention on Climate Change, "About the Secretariat," https://unfccc.int/about-us/about-the-secretariat.

13. Tim Wirth, "Interview," PBS, https://www.pbs.org/wgbh/pages/frontline/hotpolitics/interviews/wirth.html.

14. William K. Stevens, "Split Over Poorer Countries' Role Puts Cloud on Global-Warming Talks," *New York Times*, December 6, 1997, https://www.nytimes.com/1997/12/06/world/split-over-poorer-countries-role-puts-cloud-on-global-warming-talks.html.

15. Chasek, Downie, and Brown, *Global Environmental Politics*, 7th ed., 169–70.

16. William K. Stevens, "Meeting Reaches Accord to Reduce Greenhouse Gases," *New York Times*, December 11, 1997, https://www.nytimes.com/1997/12/11/world/meeting-reaches-accord-to-reduce-greenhouse-gases.html.

17. Chasek, Downie, and Brown, *Global Environmental Politics*, 7th ed., 171.

18. John H. Cushman Jr., "Talks on Global Warming Treaty Resuming Today," *New York Times*, November 2, 1998, https://www.nytimes.com/1998/11/02/world/talks-on-global-warming-treaty-resuming-today.html.

19. Stevens, "Meeting Reaches Accord to Reduce Greenhouse Gases."

20. United Nations, "United Nations Treaty Collection, Chapter XXVII, Environment. 7.a. Kyoto Protocol to the United Nations Framework Convention on Climate Change," https://treaties.un.org/Pages/ViewDetails.aspx?src=TREATY &mtdsg_no=XXVII-7-a&chapter=27&clang=_en.

21. Quoted in Chasek, Downie, and Brown, *Global Environmental Politics*, 7th ed., 171.

22. U.S. share calculated from data in Chasek, Downie, and Brown, *Global Environmental Politics*, 7th ed., 164.

23. Chasek, Downie, and Brown, *Global Environmental Politics,* 171.

24. For a fuller explanation of the various bodies created by the two treaties, see United Nations Framework Convention on Climate Change, "What Are Bodies?" (2018), https://unfccc.int/process/bodies/the-big-picture/what-are-bodies.

25. Chasek, Downie, and Brown, *Global Environmental Politics*, 174–75.

26. See John C. Berg, "Environmental Policy: The Success and Failure of Obama," in *The Obama Presidency: Promise and Performance*, ed. William Crotty (Lanham, MD: Lexington Books, 2012), 85–101.

27. Barack Obama, "A New Chapter on Climate Change," video address (2008), https://www.youtube.com/watch?v=hvG2XptIEJk; Charles F. Parker and Christer Karlsson, "The UN Climate Change Negotiations and the Role of the United States: Assessing American Leadership from Copenhagen to Paris," *Environmental Politics* 27, no. 3 (February 22, 2018): 5.

28. Quoted in Suzanne Goldenberg, "Barack Obama's U.S. Climate Change Bill Passes Key Congress Vote," *Guardian*, June 26, 2009, https://www.theguardian.com/environment/2009/jun/27/barack-obama-climate-change-bill.

29. Parker and Karlsson, "The UN Climate Change Negotiations and the Role of the United States," 5.

30. Peter Christoff, "Cold Climate in Copenhagen: China and the United States at COP15," *Environmental Politics* 19, no. 4 (2010): 638.

31. Christoff, "Cold Climate in Copenhagen," 650.

32. Christoff, "Cold Climate in Copenhagen," 640.

33. Christoff, "Cold Climate in Copenhagen," 639–40.

34. Bill McKibben, "Copenhagen: Things Fall Apart and an Uncertain Future Looms," *Yale Environment 360*, December 21, 2009, http://e360.yale.edu/feature/copenhagen_things_fall_apart_and_an_uncertain_future_looms/2225/.

35. Quoted in David Corn and Kate Sheppard, "Obama's Copenhagen Deal: How It Came About—and Why It May Not Be a Real Deal," *Mother Jones*, December 19, 2009, https://www.motherjones.com/environment/2009/12/obamas-copenhagen-deal/.

36. McKibben, "Copenhagen: Things Fall Apart and an Uncertain Future Looms."

37. Christoff, "Cold Climate in Copenhagen," 640–41.

38. Christoff, "Cold Climate in Copenhagen," 641–42.

39. Potsdam Institute for Climate Impact Research, "Ambitions of Only Two Developed Countries Sufficiently Stringent for 2°C," February 3, 2010,

https://www.pik-potsdam.de/news/in-short/archive/2010/ambition-of-only-two
-developed-countries-sufficiently-stringent-for-2b0c.

40. John C. Berg, "Obama and the Environment," in *Obama's Washington: Political Leadership in a Partisan Era*, ed. Clodagh Harrington (London: Institute of Latin American Studies, 2014), 158.

41. U.S. domestic GHG reduction efforts will be discussed in detail in chapter 7.

42. Daniel Bodansky, "The Paris Climate Change Agreement: A New Hope?," *American Journal of International Law* 110, no. 2 (2016): 292–93.

43. Bodansky, "The Paris Climate Change Agreement," 293.

44. See John C. Berg, "Obama and the Environment," in *The Obama Presidency and the Politics of Change*, eds. Edward Ashbee and John Dumbrell, Studies of the Americas (Cham: Palgrave Macmillan, 2017), 217–34.

45. Parker and Karlsson, "The UN Climate Change Negotiations and the Role of the United States," 8.

46. Radoslav S. Dimitrov, "Climate Diplomacy," in *Research Handbook on Climate Governance*, eds. Karin Bäckstrand and Eva Lövbrand (Cheltenham, England: Edward Elgar, 2015), 97–108; Radoslav S. Dimitrov, "The Paris Agreement on Climate Change: Behind Closed Doors," *Global Environmental Politics* 16, no. 3 (August 2016): 9.

47. Bodansky, "The Paris Climate Change Agreement," 293–94; Dimitrov, "The Paris Agreement on Climate Change," 6–7.

48. Bodansky, "The Paris Climate Change Agreement," 296–97.

49. Lavanya Rajamani, "Ambition and Differentiation in the 2015 Paris Agreement: Interpretive Possibilities and Underlying Politics," *International & Comparative Law Quarterly* 65, no. 2 (April 2016): 496–97.

50. Bodansky, "The Paris Climate Change Agreement," 298–300; Rajamani, "Ambition and Differentiation in the 2015 Paris Agreement," 507–13.

51. United Nations Framework Convention on Climate Change, *Paris Agreement*, https://treaties.un.org/Pages/ViewDetails.aspx?src=TREATY&mtdsg_no=XXVII-7-d&chapter=27&clang=_en.

52. Bodansky, "The Paris Climate Change Agreement," 315; signatories to the Paris Pledge for Action can be found at http://www.parispledgeforaction.org/whos-joined/.

53. Michael D. Shear, "Trump Will Withdraw U.S. from Paris Climate Agreement," *New York Times*, June 1, 2017.

54. Shear, "Trump Will Withdraw U.S. from Paris Climate Agreement."

Chapter 7

1. Walter A. Rosenbaum, *Environmental Politics and Policy,* 9th ed. (Washington, DC: CQ Press, 2014), 375.

2. Lieberman's partisan status is complex. In the 2006 election, Lieberman was defeated in the Democratic primary but ran and won as an Independent. For the next two years, he was listed as an "Independent Democrat" in the Senate

and caucused with the Democrats. He remained a registered Democrat in Connecticut.

3. Mary Clare Jalonick, "Defeat of Senate Global Warming Bill Highlights Worries over Economic Impact," *CQ Weekly*, November 1, 2003, 2709–10.

4. See chapter 4.

5. Manu Raju, "Anxious Coal Industry Deep in Emissions Debate," *CQ Weekly*, March 19, 2007, 795–96; Kara Sissell, "House Leaders Release Draft Climate Legislation," *Chemical Week*, no. 32 (2008): 10.

6. Armin Rosencranz, "U.S. Climate Change Policy under G. W. Bush," *Golden Gate University Law Review* 32, no. 4 (January 2002): 482.

7. National Energy Policy Development Group, *Reliable, Affordable, and Environmentally Sound Energy for America's Future* (Washington, DC: White House, 2001), http://www.dtic.mil/dtic/tr/fulltext/u2/a392171.pdf; Rosencranz, "U.S. Climate Change Policy under G. W. Bush," 485. See also Michael Klare, "Bush-Cheney Energy Strategy: Procuring the Rest of the World's Oil," *Foreign Policy in Focus*, January 2004, https://www.thegoldenaura.com/wp-content/uploads/2004/09/bushcheneystrategy.pdf.

8. Rosencranz, "U.S. Climate Change Policy under G. W. Bush," 488.

9. Quoted in Danny Hakim, "States Plan Suit to Prod U.S. on Global Warming," *New York Times*, October 3, 2003, https://www.nytimes.com/2003/10/04/business/states-plan-suit-to-prod-us-on-global-warming.html.

10. Rosemary O'Leary, "Environmental Policy in the Courts," in *Environmental Policy: New Directions for the Twenty-First Century,* 10th ed., eds. Norman J. Vig and Michael E. Kraft (Thousand Oaks, CA: CQ Press, 2019), 150; Hakim, "States Plan Suit."

11. O'Leary, "Environmental Policy in the Courts," 150–51; John C. Berg, "How Far Can He Go? Prospects for Trump to Reverse Obama's Climate Policies," paper presented at the annual conference of the American Politics Group, University of Oxford, January 4–6, 2018, 2.

12. "Promises about Environment on the Obameter," Politifact, http://www.politifact.com/truth-o-meter/promises/obameter/subjects/environment/; Berg, "Environmental Policy," 86–87.

13. Steven Cohen, "The Transition from Environmental Politics to Sustainability Politics," *Huffington Post*, January 31, 2011, http://www.huffingtonpost.com/steven-cohen/the-transition-from-envir_b_816198.html; Berg, "Environmental Policy," 96–97.

14. Berg, "Obama and the Environment," 227–29.

15. Center for Responsive Politics, "Coal Mining: Top Recipients, 2008 Presidential Candidates," OpenSecrets.org, https://www.opensecrets.org/industries/recips.php?ind=E1210&cycle=2008&recipdetail=P&mem=N&sortorder=U.

16. Coral Davenport, "No Senate Sequel for Influential Climate Group," *CQ Weekly*, March 29, 2010, 735–36.

17. "Highlights of the House Climate Bill," *CQ Weekly*, June 29, 2009, 1517; "For the Record: How the House Voted June 26," *CQ Weekly*, July 6, 2009, 1588.

18. Calculated by the author from data in Farhana Hossain et al., "The Stimulus Plan: How to Spend $787 Billion," *New York Times*, February 17, 2009, https://www.nytimes.com/interactive/projects/44th_president/stimulus#. It is likely that additional climate spending was contained in parts of the stimulus not labeled as such.

19. Cohen, "Transition from Environmental Politics to Sustainability Politics."

20. Barack Obama, *Remarks of President Barak Obama in State of the Union Address—as Prepared for Delivery* (Washington, DC: White House, Office of the Press Secretary, 2011), http://www.whitehouse.gov/the-press-office/2011/01/25/remarks-president-barack-obama-state-union-address-prepared-delivery.

21. Doug Pibel and Van Jones, "Van Jones: Why I'm Going to Washington," *Yes!*, March 10, 2009, https://www.yesmagazine.org/issues/food-for-everyone/van-jones-why-i2019m-going-to-washington.

22. Cohen, "Transition from Environmental Politics to Sustainability Politics."

23. George W. Bush, "Cooperation among Agencies in Protecting the Environment with Respect to Greenhouse Gas Emissions from Motor Vehicles, Nonroad Vehicles, and Nonroad Engines," Exec. Order No. 13432, *Federal Register* 72, no. 94 (May 16, 2007): 27717–19, https://www.govinfo.gov/app/details/FR-2007-05-16/07-2462.

24. Environmental Protection Agency, *EPA's Endangerment Finding: Legal Background*, December 7, 2009, https://www.epa.gov/sites/production/files/2016–08/documents/endangermentfinding_legalbasis.pdf.

25. Environmental Protection Agency, *EPA's Endangerment Finding: Legal Background*; Environmental Defense Fund, "Overview of EPA Endangerment Finding," released September 2001, https://www.edf.org/overview-epa-endangerment -finding. The finding is 40 C.F.R. Chapter 1, *Federal Register* 74, no. 239 (December 15, 2009): 66495–546, and the technical support document can be found at https://www.epa.gov/sites/production/files/2016-08/documents/endangerment _tsd.pdf.

26. U.S. Environmental Protection Agency, "Endangerment and Cause or Contribute Findings for Greenhouse Gases under the Section 202(a) of the Clean Air Act," https://www.epa.gov/ghgemissions/endangerment-and-cause-or-con tribute-findings-greenhouse-gases-under-section-202a-clean.

27. Robert A. Levy, "Who Elected Lisa Jackson?," *CATO Policy Report* 33, no. 2 (March-April 2011): 2.

28. The video can be seen at http://www.postcarbon.org/video/51255-mc kibben-first-take-on-climate#.

29. McKibben, "Copenhagen: Things Fall Apart and an Uncertain Future Looms."

30. Ted Nordhaus and Michael Shellenberger, *Break Through: From the Death of Environmentalism to the Politics of Possibility* (Boston: Houghton Mifflin, 2007).

31. The next several paragraphs are adapted from Berg, "Leave It in the Ground."

32. James Hansen et al., "Target Atmospheric CO_2: Where Should Humanity Aim?," *Open Atmospheric Science Journal* 2, no. 1 (December 2008): 228.

33. Thomas Stocker et al., *Climate Change 2013: The Physical Science Basis, Contribution of Working Group I to the Fifth Assessment Report of the Intergovernmental Panel on Climate Change* (Cambridge, MA: Cambridge University Press, 2013), 25.

34. Stocker et al., *Climate Change 2013: The Physical Science Basis*, 26.

35. Bill McKibben, "The Reckoning," *Rolling Stone*, no. 1162 (August 2, 2012): 52–60.

36. See Berg, "Cape Wind."

37. "Keystone XL Pipeline Overview," *Congressional Digest* 90, no. 10 (December 2011): 290.

38. "James Hansen: Taking Heat for Decades," *Bulletin of the Atomic Scientists* 69, no. 4 (July 2013): 4.

39. "Bill McKibben: Actions Speak Louder Than Words," *Bulletin of the Atomic Scientists* 68, no. 2 (March 2012): 7.

40. Jordan Michael Smith, "Northern Promises," *World Affairs* 176, no. 2 (July-August 2013): 77.

41. Andy Woloszyn and David Finkel, "Rallying to Stop Keystone XL: Tipping Point for a Movement?," *Against the Current* 28, no. 1 (March-April 2013): 7.

42. President Trump has since resuscitated the pipeline.

43. Mark Hertsgaard, "Emissions Impossible," *Mother Jones* 37, no. 3 (May-June 2012): 42.

44. Quoted in Hertsgaard, "Emissions Impossible," 38.

45. Sierra Club, "2013: A Landmark Year for Clean Energy; Twilight for Coal," *Compass: Pointing the Way to a Clean Energy Future*, December 12, 2013, http://sierraclub.typepad.com/compass/2013/12/2013-a-landmark-year-for-clean-energy-twilight-for-coal-.html.

46. Sierra Club, "Victories," https://content.sierraclub.org/coal/victories.

47. Hertsgaard, "Emissions Impossible."

48. Scott Parkin of the Rainforest Action Network, quoted in Robert S. Eshelman, "Cracking Big Coal," *Nation* 290, no. 17 (May 3, 2010): 18.

49. Eshelman, "Cracking Big Coal," 18.

50. Rick Reed of the Garfield Foundation, quoted in Hertsgaard, "Emissions Impossible," 42.

51. Science note: Coal is basically solid carbon with a few impurities (and those tend to be highly polluting substances such as sulfur and mercury); all the energy produced comes from the burning of carbon and produces CO_2 as the waste product. Petroleum and natural gas are hydrocarbons; some of the energy produced in combustion comes from the carbon, but some comes from hydrogen. The waste products are therefore both CO_2 and H_2O. While water is also a greenhouse gas, the water vapor in the atmosphere is in equilibrium with the vast stock of water in the ocean, so that the concentration of H_2O in the atmosphere is determined by temperature and pressure, not by emissions.

52. Barack Obama, *Remarks by the President in the State of the Union Address*, (Washington, DC: White House, Office of the Press Secretary, 2013),

https://obamawhitehouse.archives.gov/the-press-office/2013/02/12/remarks
-president-state-union-address.

53. *U.S.-China Joint Announcement on Climate Change*, press statement (Washington, DC: White House, Office of the Press Secretary, 2014), https://obamawhitehouse.archives.gov/the-press-office/2014/11/11/us-china-joint-announcement-climate-change. The agreement was announced in Beijing on November 12, but the press release is dated November 11 due to the international time difference.

54. John Kerry, *Keystone XL Pipeline Permit Determination*, (Washington, DC: Department of State, 2015), https://2009–2017.state.gov/secretary/remarks/2015/11/249249.htm.

55. David M. Konisky and Neal D. Woods, "Environmental Policy, Federalism, and the Obama Presidency," *Publius: The Journal of Federalism* 46, no. 3 (Summer 2016): 372.

56. David R. Mayhew, *Divided We Govern: Party Control, Lawmaking, and Investigations, 1946–2002* (New Haven, CT: Yale University Press, 2005).

57. Mann and Ornstein, *It's Even Worse Than It Looks*.

58. Andrew Rudalevige, "The Contemporary Presidency: The Obama Administrative Presidency: Some Late-Term Patterns," *Presidential Studies Quarterly* 46, no. 4 (December 2016): 870.

59. Barack Obama, *Remarks by the President in Announcing the Clean Power Plan* (Washington, DC: White House, Office of the Press Secretary, 2015), https://obamawhitehouse.archives.gov/the-press-office/2015/08/03/remarks-president-announcing-clean-power-plan. "Gina" refers to Gina McCarthy, head of the EPA from 2013 to 2017.

60. Richard N. L. Andrews, "The Environmental Protection Agency," in *Environmental Policy: New Directions for the Twenty-First Century,* 10th ed., eds. Norman J. Vig and Michael E. Kraft (Thousand Oaks, CA: CQ Press, 2019), 180.

61. Jonathan H. Adler, "Heat Expands All Things: The Proliferation of Greenhouse Gas Regulation under the Obama Administration," *Harvard Journal of Law & Public Policy* 34, no. 2 (Spring 2011): 435.

62. Adam Liptak and Coral Davenport, "Supreme Court Deals Blow to Obama's Efforts to Regulate Coal Emissions," *New York Times*, February 9, 2016. https://www.nytimes.com/2016/02/10/us/politics/supreme-court-blocks-obama-epa-coal-emissions-regulations.html; Coral Davenport, "Obama Climate Plan, Now in Court, May Hinge on Error in 1990 Law," *New York Times*, September 25, 2016, https://www.nytimes.com/2016/09/26/us/politics/obama-court-clean-power-plan.html.

Chapter 8

1. Leigh Raymond, *Reclaiming the Atmospheric Commons: The Regional Greenhouse Gas Initiative and a New Model of Emissions Trading* (Cambridge, MA: MIT Press, 2016); Jonathan London et al., "Racing Climate Change: Collaboration and Conflict in California's Global Climate Change Policy Arena," *Global Environmental Change* 23, no. 4 (2013): 791–99.

2. Trump's tweet can be found at https://twitter.com/realDonaldTrump/status/265895292191248385.

3. Justin Worland, "Donald Trump Does Not Believe in Man-Made Climate Change, Campaign Manager Says," *Time*, September 27, 2016, http://time.com/4509488/donald-trump-climate-change-hoax/.

4. Trump's remarks to the *Times* are quoted in Philip Bump, "What's Donald Trump's Position on Climate Change? All of Them," *Washington Post*, November 22, 2016, https://www.washingtonpost.com/news/the-fix/wp/2016/11/22/whats-donald-trumps-position-on-climate-change-all-of-them.

5. However, Trump might have won these states anyway, as he won such non-coal-producing states as Michigan and Wisconsin.

6. Quoted in Lisa Friedman, "How a Coal Baron's Wish List Became President Trump's To-Do List," *New York Times*, January 9, 2018, https://www.nytimes.com/2018/01/09/climate/coal-murray-trump-memo.html.

7. Logan Layden, "Court Losses Won't Deter Attorney General Scott Pruitt in His Fight with the EPA," *StateImpact Oklahoma*, June 12, 2014, https://stateimpact.npr.org/oklahoma/2014/06/12/court-losses-wont-deter-attorney-general-scott-pruitt-in-his-fight-with-the-epa/; Paul Monies, "Oklahoma Attorney General, 11 Others File Lawsuit against EPA Over 'Sue and Settle' Tactics," *NewsOK*, July 17, 2013, https://newsok.com/article/3862959/oklahoma-attorney-general-11-others-file-lawsuit-against-epa-over-sue-and-settle-tactics.

8. Coral Davenport, "EPA Chief Doubts Consensus View of Climate Change," *New York Times*, March 9, 2017, https://www.nytimes.com/2017/03/09/us/politics/epa-scott-pruitt-global-warming.html.

9. Quoted in Dino Grandoni, Dennis Brady, and Chris Mooney, "EPA's Scott Pruitt Asks Whether Global Warming 'Necessarily Is a Bad Thing,'" *Washington Post*, February 8, 2018, https://www.washingtonpost.com/news/energy-environment/wp/2018/02/07/scott-pruitt-asks-if-global-warming-necessarily-is-a-bad-thing.

10. "Scott Pruitt: Controversial Trump Environment Nominee Sworn In," *BBC News*, February 17, 2017, https://www.bbc.com/news/world-us-canada-39010374.

11. Sam Wolfson, "The Ethics Scandals That Eventually Forced Scott Pruitt to Resign," *Guardian*, July 5, 2018, https://www.theguardian.com/us-news/2018/jul/05/scott-pruitt-what-it-took-to-get-him-to-resign-from-his-epa-job; Eli Watkins and Clare Foran, "EPA Chief Scott Pruitt's Long List of Controversies," *CNN Politics*, July 5, 2018, https://www.cnn.com/2018/04/06/politics/scott-pruitt-controversies-list/index.html.

12. Lisa Friedman, "Andrew Wheeler, New EPA Chief, Details His Energy Lobbying Past," *New York Times*, August 1, 2018, https://www.nytimes.com/2018/08/01/climate/andrew-wheeler-epa-lobbying.html.

13. Lisa Friedman, "Trump Says He'll Nominate Andrew Wheeler to Head the EPA," *New York Times*, November 16, 2018, https://www.nytimes.com/2018/11/16/climate/trump-andrew-wheeler-epa.html.

14. Quoted in Tim Murphy, "Trump's Interior Nominee Was for Climate Action before He Was against It," *Mother Jones*, December 14, 2016, https://www.motherjones.com/politics/2016/12/ryan-zinke-donald-trump-climate-change/.

15. Ryan Zinke and Scott McEwen, *American Commander: Serving a Country Worth Fighting for and Training the Brave Soldiers Who Lead the Way* (Nashville: Thomas Nelson, 2016).

16. Tom Lutey, "Zinke Resigns Delegate Post over Public Lands Disagreement; Still Will Speak at RNC," *Billings Gazette*, July 15, 2016, https://billingsgazette .com/news/local/zinke-resigns-delegate-post-over-public-lands-disagreement- still-will/article_8109f084-d199–50dd-b223–9fd3557a738d.html.

17. Wes Siler, "Ryan Zinke Has Fired the DOI Inspector General," *Outside*, October 15, 2018, https://www.outsideonline.com/2355936/zinke-fires-inspector -general.

18. Umair Irfan, "Interior Secretary Ryan Zinke Might Face a Criminal Inves- tigation," *Vox*, November 5, 2018, https://www.vox.com/2018/10/31/18044860/ ryan-zinke-interior-investigation-ethics-justice.

19. Carl Hofacker, "Interior Secretary Ryan Zinke Accuses Democrat Who Wants Him to Resign of 'Drunken and Hostile Behavior,'" *USA Today*, November 30, 2018, https://www.usatoday.com/story/news/politics/2018/11/30/ryan-zinke -accuses-rep-paul-grijalva-drunken-hostile-behavior/2163291002/.

20. Juliet Eilperin and Josh Dawsey, "Ryan Zinke Resigns as Interior Secretary amid Multiple Investigations," *Chicago Tribune*, December 15, 2018, https://www .chicagotribune.com/news/nationworld/politics/ct-ryan-zinke-resigns-interior -secretary-20181215-story.html.

21. Lydia Wheeler, "Meet the Powerful Group behind Trump's Judicial Nomi- nations," *The Hill*, November 16, 2017, https://thehill.com/regulation/court -battles/360598-meet-the-powerful-group-behind-trumps-judicial-nominations; Jason Zengerie, "How the Trump Administration Is Remaking the Courts," *New York Times*, August 22, 2018, https://www.nytimes.com/2018/08/22/magazine/ trump-remaking-courts-judiciary.html.

22. Quoted in Shear, "Trump Will Withdraw U.S. from Paris Climate Agreement."

23. Shear, "Trump Will Withdraw U.S. from Paris Climate Agreement."

24. Banks's resignation had not been for policy reasons but because he had been denied a full security clearance due to marijuana use when he was a youth. See Lisa Friedman, "Former Trump Aide Calls Paris Climate Accord 'A Good Republican Agreement,'" *New York Times*, February 22, 2018, https://www .nytimes.com/2018/02/22/climate/george-david-banks.html.

25. Intergovernmental Panel on Climate Change, *Special Report: Global Warm- ing of 1.5°C* (2018), https://www.ipcc.ch/sr15/.

26. The quotation is from Elliot Diringer of the Center for Climate and Energy Solutions; see Frank Jordans, "Talks Adopt 'Rulebook' to Put Paris Climate Deal into Action," *New York Times*, December 16, 2018, https://www.apnews.com/92d a6d165f6c4addbbbb82ac7a3eed84.

27. Lisa Friedman and Brad Plumer, "EPA Announces Repeal of Major Obama- Era Carbon Emissions Rule," *New York Times*, October 9, 2017, https://www .nytimes.com/2017/10/09/climate/clean-power-plan.html.

28. Environmental Protection Agency, "State Guidelines for Greenhouse Gas Emissions from Existing Electric Utility Generating Units: Advance Notice of

Proposed Rulemaking," *Federal Register* 82, no. 248 (December 28, 2017): 61507–19, https://www.govinfo.gov/content/pkg/FR-2017-12-28/pdf/2017-27793.pdf.

29. Environmental Protection Agency, "Repeal of Carbon Pollution Emission Guidelines for Existing Stationary Sources: Electric Utility Generating Units: Proposed Rule," *Federal Register* 82, no. 198 (October 16, 2017): 48039, https://www.govinfo.gov/content/pkg/FR-2017-12-28/pdf/2017-27793.pdf.

30. The steps in the process are laid out on the EPA webpage "Electric Utility Generating Units: Repealing the Clean Power Plan," https://www.epa.gov/stationary-sources-air-pollution/electric-utility-generating-units-repealing-clean-power-plan.

31. Environmental Protection Administration, "Proposal: Affordable Clean Energy (ACE) Rule," August 21, 2018, https://www.epa.gov/stationary-sources-air-pollution/proposal-affordable-clean-energy-ace-rule; Kallanish Energy, "Trump's EPA Releases Affordable Clean Energy Rule," August 22, 2018, http://www.kallanishenergy.com/2018/08/22/trumps-epa-releases-affordable-clean-energy-rule/.

32. All quotations from Timothy Cama, "Trump's EPA Budget Cuts Hit Strong Opposition at House Panel," *The Hill*, June 15, 2017, https://thehill.com/policy/energy-environment/337979-trumps-epa-budget-cuts-hit-strong-opposition-at-house-panel. See also "Even Some GOP Lawmakers Don't Like Trump's Cuts to Environmental Programs," *CQ Magazine*, June 12, 2017.

33. Devin Henry, "House Bill Would Cut EPA Funding by $528M," *The Hill*, July 11, 2017, https://thehill.com/policy/energy-environment/341507-house-bill-would-cut-epa-funding-by-528m.

34. Colin Staub, "EPA's Recycling Funding Untouched in Final Budget," *Resource Recycling*, March 27, 2018, https://resource-recycling.com/recycling/2018/03/27/epas-recycling-funding-untouched-in-final-budget/.

35. *CQ Magazine,* "2019 Budget: Environmental Protection Agency,", March 12, 2018;, *CQ Magazine*, "2019 Appropriations: Interior-Environment,", July 30, 2018.

36. Jesse Byrnes, "EPA Begins Offering Buyouts to Cut Staff: Report," *The Hill*, June 1, 2017, https://thehill.com/homenews/administration/336023-epa-begins-offering-buyouts-to-cut-staff-report; Douglas Main, "EPA, Interior Plan to Cut More Than 5,000 Staff by 2018," *Newsweek*, June 21, 2017, https://www.newsweek.com/epa-interior-plan-cut-more-5000-staff-2018–627998; Louis Jacobson, "How Much Have Republicans Cut EPA Budget, Staff in Two Years?," *Politifact*, April 26, 2018, https://www.politifact.com/truth-o-meter/statements/2018/apr/26/evan-jenkins/how-much-have-republicans-cut-epa-budget-staff-two/; U.S. Office of Personnel Management, "FedScope: Employment Cubes (September 2018)," posted March 19, 2019, https://www.fedscope.opm.gov/employment.asp.

37. "Global Covenant of Mayors for Climate & Energy's Cities Have Committed to Reducing Emissions Equal to Removing All U.S. Cars off the Road," press

release, September 10, 2018, https://www.globalcovenantofmayors.org/wp-con
tent/uploads/2018/09/GCoM-Data-Press-Release_GCAS_.pdf. See also "About
the Global Covenant of Mayors for Climate & Energy," https://www.globalcove
nantofmayors.org/about/.

38. U.S. Conference of Mayors, "Mayors Climate Protection Agreement," Mayors Climate Protection Center, https://www.usmayors.org/mayors-climate-protection
-center/.

39. *U.S. Mayors Report on a Decade of Global Climate Leadership: Selected Mayor
Profiles* (Washington, DC: U.S. Conference of Mayors, December 2015), http://
www.usmayors.org/wp-content/uploads/2017/06/1205-report-climateaction
.pdf.

40. Although the city is almost two thousand years old, the modern office of
mayor was created in 2000.

41. John Vidal, "Plane Speaking," *Guardian*, November 1, 2006, https://www
.theguardian.com/environment/2006/nov/01/travelsenvironmentalimpact.local
government; David Adam and Hugh Muir, "Cleaning Up the Big Smoke: Livingstone Plans to Cut Carbon Emissions by 60%," *Guardian*, February 27, 2007,
https://www.theguardian.com/environment/2007/feb/27/energy.localgovernment.

42. Lucy Flynn, "Boris Johnson's Decision Not to Divest Goes against His
Climate Goals for London," *Guardian*, May 21, 2015, https://www.theguardian.
com/environment/2015/may/21/boris-johnsons-decision-not-to-divest-goes
-against-his-climate-goals-for-london.

43. "80×50" refers to a goal of 80 percent GHG reduction by 2050. Climate
Action Plan Steering Committee, *Greenovate Boston: 2014 Climate Action Plan
Update* (Boston: City of Boston, 2014), 72.

44. City of Boston, "Fiscal Year 2019 Budget: Capital Projects, Environment,"
229, https://www.boston.gov/sites/default/files/imce-uploads/2019-04/v2_06-
_19_a_environment-energy-and-open-space-cabinet.pdf/.

45. Chicago Climate Task Force, *Chicago Climate Action Plan: Our City, Our
Future* (Chicago: City of Chicago, 2008), 12–43, http://www.chicagoclimateac-
tion.org/filebin/pdf/finalreport/CCAPREPORTFINALv2.pdf.

46. Julia Parzen, *Lessons Learned: Creating the Chicago Climate Action Plan* (Chicago: Chicago Climate Action Plan, 2009), http://www.chicagoclimateaction.
org/filebin/pdf/LessonsLearned.pdf.

47. Hugh Bartling, "Climate Policy and Leadership in a Metropolitan Region:
Cases from the United States," *Local Economy* 32, no. 4 (June 2017): 345.

48. City of Chicago, *Sustainable Chicago 2015: Action Agenda* (Chicago: City of
Chicago, 2012), 32–35, https://www.chicago.gov/content/dam/city/progs/env/
SustainableChicago2015.pdf.

49. Rahm Emmanuel, *Statement from Mayor Emmanuel on Clean Power Plan*,
press release (Chicago: City of Chicago, Office of the Mayor, 2017), https://www
.chicago.gov/city/en/depts/mayor/press_room/press_releases/2017/october/
CleanPowerPlan.html.

50. Data from International Monetary Fund, "World Economic and Financial Surveys: World Economic Database," https://www.imf.org/external/pubs/ft/weo/2017/02/weodata/index.aspx.

51. California Air Resources Board, "AB 32 Scoping Plan," page last reviewed January 8, 2018, https://www.arb.ca.gov/cc/scopingplan/scopingplan.htm.

52. California Air Resources Board, *California's 2017 Climate Change Scoping Plan: Executive Summary* (Sacramento: California Air Resources Board, 2017), https://www.arb.ca.gov/cc/scopingplan/scoping_plan_2017_es.pdf.

53. London et al., "Racing Climate Change."

54. "What is RGGI," Stronger RGGI for a Clean Energy Economy, https://www.cleanenergyeconomy.us/#about-rggi.

55. David Abel, "Nine States Aggressively Step Up Plans to Cut Emissions," *Boston Globe*, August 23, 2017, https://www.bostonglobe.com/metro/2017/08/23/emissions/Bu8VXv5jqkqldVNKXpafON/story.html.

56. Raymond, *Reclaiming the Atmospheric Commons*.

57. Michael D. Shear and Alison Smale, "Leaders Lament U.S. Withdrawal, but Say It Won't Stop Climate Efforts," *New York Times*, June 2, 2017, https://www.nytimes.com/2017/06/02/climate/paris-climate-agreement-trump.html.

58. Somini Sengupta et al., "As Trump Exits Paris Agreement, Other Nations Are Defiant," *New York Times*, June 1, 2017, https://www.nytimes.com/2017/06/01/world/europe/climate-paris-agreement-trump-china.html.

59. Brad Plumer, "Climate Negotiators Reach an Overtime Deal to Keep Paris Pact Alive," *New York Times*, December 15, 2018, https://www.nytimes.com/2018/12/15/climate/cop24-katowice-climate-summit.html.

60. Sierra Club, "About Us," Beyond Coal, https://content.sierraclub.org/coal/about-the-campaign.

61. Rhodium Group Energy and Climate Staff, "Preliminary U.S. Emissions Estimates for 2018," January 8, 2019, https://rhg.com/research/preliminary-us-emissions-estimates-for-2018/.

62. Robert Hockett, "The Green New Deal Is What Our Planet Has Been Waiting For," *The Hill*, February 7, 2019.

63. Gregory Krieg, "Who Is Alexandria Ocasio-Cortez?," *CNN Politics*, June 27, 2018, https://www.cnn.com/2018/06/27/politics/who-is-alexandria-ocasio-cortez/index.html.

64. Elvina Nawaguna, "Democrats Unveil Green New Deal That Would Push Government to Make Radical Changes," *Roll Call*, February 7, 2019, https://www.rollcall.com/news/congress/democrats-offer-green-new-deal-resolution-for-economic-overhaul.

65. Pelosi statements quoted in Owen Daugherty, "Pelosi Praises Enthusiasm Behind 'Green New Deal' After Seeming to Brush It Off," *The Hill*, February 7, 2019, https://thehill.com/policy/energy-environment/428956-pelosi-praises-enthusiasm-behind-green-new-deal-after-seeming-to-brush-it-off.

66. At the time of writing, the resolution had not yet been assigned a number. A copy of the final draft may be found online, https://assets.documentcloud.org/documents/5729033/Green-New-Deal-FINAL.pdf.

Bibliography

Abel, David. "Nine States Aggressively Step Up Plans to Cut Emissions." *Boston Globe*, August 23, 2017. https://www.bostonglobe.com/metro/2017/08/23/emissions/Bu8VXv5jqkqldVNKXpafON/story.html.

Ackerman, Bruce A., and William T. Hassler. *Clean Coal/Dirty Air: or, How the Clean Air Act Became a Multibillion-Dollar Bail-Out for High-Sulfur Coal Producers and What Should Be Done About It.* New Haven: Yale University Press, 1981.

Adam, David, and Hugh Muir. "Cleaning Up the Big Smoke: Livingstone Plans to Cut Carbon Emissions by 60%." *Guardian*, February 27, 2007. https://www.theguardian.com/environment/2007/feb/27/energy.localgovernment.

Adams, Sean Patrick. "The US Coal Industry in the Nineteenth Century." EH.net Encyclopedia. Edited by Robert Whaples, January 23, 2003. http://eh.net/encyclopedia/the-us-coal-industry-in-the-nineteenth-century.

Adler, Jonathan H. "Heat Expands All Things: The Proliferation of Greenhouse Gas Regulation under the Obama Administration." *Harvard Journal of Law & Public Policy* 34, no. 2 (Spring 2011): 421–52.

Alcott, Blake. "Jevons' Paradox." *Ecological Economics* 54, no. 1 (July 1, 2005): 9–21.

Alinsky, Saul. *John L. Lewis: An Unauthorized Biography.* New York: Putnam, 1949.

Andrews, Richard N. L. "The Environmental Protection Agency." In *Environmental Policy: New Directions for the Twenty-First Century.* 10th ed., edited by Norman J. Vig and Michael E. Kraft, 168–92. Thousand Oaks, CA: CQ Press, 2019.

Andrews, Thomas G. *Killing for Coal: America's Deadliest Labor War.* Cambridge, MA: Harvard University Press, 2008.

Austen, Ian. "Technology to Make Clean Energy from Coal Is Stumbling in Practice." *New York Times*, March 29, 2016. https://www.nytimes.com/2016/03/30/business/energy-environment/technology-to-make-clean-energy-from-coal-is-stumbling-in-practice.html.

Bakalaki, Alexandra. "Chemtrails, Crisis, and Loss in an Interconnected World." *Visual Anthropology Review* 32, no. 1 (May 2016): 12–23.

Bari, Judi. *Timber Wars.* Monroe, ME: Common Courage, 1994.

Bartling, Hugh. "Climate Policy and Leadership in a Metropolitan Region: Cases from the United States." *Local Economy* 32, no. 4 (June 2017): 336–51.

Bell, Michelle L., and Devra Lee Davis. "Reassessment of the Lethal London Fog of 1952: Novel Indicators of Acute and Chronic Consequences of Acute Exposure to Air Pollution." *Environmental Health Perspectives* 109, no. S3 (June 2001): S389–94.

Benson, Paul R. *The Supreme Court and the Commerce Clause, 1937–1970.* With a foreword by Maurice G. Baxter. New York: Dunellen, 1970.

Berg, John C. "Cape Wind: A Case Study in the Politics of Technology Choice." In *Beyond the Global Village: Environmental Challenges Inspiring Global Citizenship,* edited by Rafaela C. Hillerbrand and Rasmus Karlsson, 65–74. Oxford: Inter-Disciplinary Press, 2007.

Berg, John C. "Environmental Policy: The Success and Failure of Obama." In *The Obama Presidency: Promise and Performance,* edited by William Crotty, 85–101. Lanham, MD: Lexington Books, 2012.

Berg, John C. "How Far Can He Go? Prospects for Trump to Reverse Obama's Climate Policies." Paper presented at the annual conference of the American Politics Group, University of Oxford, January 4–6, 2018.

Berg, John C. "Leave It in the Ground: Science, Politics, and the Movement to End Coal Use." Paper presented at the annual conference of the American Politics Group, University of Oxford, January 5–7, 2014. http://papers .ssrn.com/sol3/papers.cfm?abstract_id=2375370.

Berg, John C. "Obama and the Environment." In *Obama's Washington: Political Leadership in a Partisan Era,* edited by Clodagh Harrington. London: Institute of Latin American Studies, 2014.

Berg, John C. "Obama and the Environment." In *The Obama Presidency and the Politics of Change,* edited by Edward Ashbee and John Dumbrell. Studies of the Americas, 217–34. Cham, Switzerland: Palgrave Macmillan, 2017.

Berg, John C., ed. *Teamsters and Turtles? U.S. Progressive Political Movements in the 21st Century.* Lanham, MD: Rowman & Littlefield, 2003.

Biello, David. "The Carbon Capture Fallacy." *Scientific American* 314, no. 1 (January 2016): 58–65.

"Bill McKibben: Actions Speak Louder Than Words." *Bulletin of the Atomic Scientists* 68, no. 2 (March 2012): 1–8.

Bodansky, Daniel. "The Paris Climate Change Agreement: A New Hope?" *American Journal of International Law* 110, no. 2 (2016), 288–319.

Boden, Tom, Bob Andres, and Gregg Marland. *Global CO_2 Emissions from Fossil-Fuel Burning, Cement Manufacture, and Gas Flaring: 1751–2014.* Washington, DC: Oak Ridge National Laboratory, 2017. http://cdiac.ornl .gov/ftp/ndp030/global.1751_2014.ems.

Bodenhamer, David J. *The Revolutionary Constitution*. New York: Oxford University Press, 2012.

Boyer, Richard O., and Herbert M. Morais. *Labor's Untold Story*. 1955. Reprint, New York: United Electrical Workers, 1970.

Brecher, Jeremy. "Stormy Weather: Climate Change and a Divided Labor Movement." *New Labor Forum* 22, no. 1 (2013): 75–81.

Brienes, Marvin. "The Fight against Smog in Los Angeles, 1943–1957." PhD diss., University of California Davis, 1975.

Brimblecombe, Peter. *The Big Smoke: A History of Air Pollution in London since Medieval Times*. London: Methuen, 1987.

Bryner, Gary C. *Blue Skies, Green Politics: The Clean Air Act of 1990 and Its Implementation*. Washington, DC: CQ Press, 1995.

Bump, Philip. "What's Donald Trump's Position on Climate Change? All of Them." *Washington Post*, November 22, 2016. https://www.washington post.com/news/the-fix/wp/2016/11/22/whats-donald-trumps-position -on-climate-change-all-of-them.

Burns, William C. G. "Geoengineering the Climate: An Overview of Solar Radiation Management Options." *Tulsa Law Review* 46, no. 2 (Winter 2010): 283–304.

Bush, George W. "Cooperation among Agencies in Protecting the Environment with Respect to Greenhouse Gas Emissions from Motor Vehicles, Nonroad Vehicles, and Nonroad Engines." Exec. Order No.13432. *Federal Register* 72, no. 94 (May 16, 2007): 27717–19. https://www.govinfo.gov/content/pkg/FR-2007-05-16/pdf/07-2462.pdf.

Byrnes, Jesse. "EPA Begins Offering Buyouts to Cut Staff: Report." *The Hill*, June 1, 2017. https://thehill.com/homenews/administration/336023-epa-begins -offering-buyouts-to-cut-staff-report.

Cairns, Rose. "Climates of Suspicion: 'Chemtrail' Conspiracy Narratives and the International Politics of Geoengineering." *Geographical Journal* 182, no. 1 (March 2016): 70–84.

Caldeira, Ken, and Michael E. Wickett. "Oceanography: Anthropogenic Carbon and Ocean pH." *Nature* 425, no. 6956 (September 25, 2003): 365.

California Air Resources Board. "AB 32 Scoping Plan." Page last reviewed January 8, 2018. https://www.arb.ca.gov/cc/scopingplan/scopingplan.htm.

California Air Resources Board. *California's 2017 Climate Change Scoping Plan: Executive Summary*. Sacramento: California Air Resources Board, 2017. https://www.arb.ca.gov/cc/scopingplan/scoping_plan_2017_es.pdf.

Cama, Timothy. "Trump's EPA Budget Cuts Hit Strong Opposition at House Panel." *The Hill*, June 15, 2017. https://thehill.com/policy/energy-environ ment/337979-trumps-epa-budget-cuts-hit-strong-opposition-at -house-panel.

Cardwell, Diane. "The Murky Future of Nuclear Power in the United States." *New York Times*, February 18, 2017. https://www.nytimes.com/2017/02/

18/business/energy-environment/nuclear-power-westinghouse-toshiba
.html.

Caudill, Harry M. *Night Comes to the Cumberlands: A Biography of a Depressed Area*. Boston: Little, Brown, 1962.

Center for Responsive Politics. "Coal Mining" OpenSecrets.org. https://www
.opensecrets.org/industries/recips.php?ind=E1210&cycle=2008&recipd
etail=P&mem=N&sortorder=U.

Center for Responsive Politics. "Coal Mining: Long-Term Contribution Trends." OpenSecrets.org. http://www.opensecrets.org/industries/totals.php?cycle
=2018&ind=E1210.

Chasek, Pamela S., David L. Downie, and Janet Welsh Brown. *Global Environmental Politics*. 7th ed. *Dilemmas in World Politics*. Boulder: Westview Press, 2017.

Chicago Climate Task Force. *Chicago Climate Action Plan: Our City, Our Future*. Chicago: City of Chicago, 2008. http://www.chicagoclimateaction.org/
filebin/pdf/finalreport/CCAPREPORTFINALv2.pdf.

Chou, Ming-Dah, Richard S. Lindzen, and Arthur Y. Hou. "Comments on 'The Iris Hypothesis: A Negative or Positive Cloud Feedback?'" *Journal of Climate* 15, no. 18 (September 15, 2002): 2713–15.

Christensen, Clayton M. *The Innovator's Dilemma: When New Technologies Cause Great Firms to Fail*. New York: Harper Business, 2000.

Christensen, Clayton M., and Michael E. Raynor. *The Innovators Solution: Creating and Sustaining Successful Growth*. Boston: Harvard Business School, 2003.

Christoff, Peter. "Cold Climate in Copenhagen: China and the United States at COP15." *Environmental Politics* 19, no. 4 (2010): 637–56.

Churchill, Ward. "From the Pinkertons to the PATRIOT Act: The Trajectory of Political Policing in the United States, 1870 to the Present." *CR: The New Centennial Review* 4, no. 1 (2004): 1–72.

Cillizza, Chris. "Trumka Hopes to Mend the AFL-CIO." *Washington Post*, July 13, 2009.

City of Boston. "Fiscal Year 2019 Budget: Capital Projects, Environment." 229. https://www.boston.gov/sites/default/files/imce-uploads/2019-04/v2
_06-_19_a_environment-energy-and-open-space-cabinet.pdf/.

City of Chicago. *Sustainable Chicago 2015: Action Agenda*. Chicago: City of Chicago, 2012. https://www.chicago.gov/content/dam/city/progs/env/Sustainable
Chicago2015.pdf.

Clarke, Robin, and Lloyd Timberlake. *Stockholm Plus Ten: Promises, Promises? The Decade since the 1972 UN Environment Conference*. Earthscan Paperback. London: International Institute for Environment and Development, 1982.

Climate Action Plan Steering Committee. *Greenovate Boston: 2014 Climate Action Plan Update*. Boston: City of Boston, 2014.

Cohen, Steven. "The Transition from Environmental Politics to Sustainability Politics." *Huffington Post*, January 31, 2011. http://www.huffingtonpost
.com/steven-cohen/the-transition-from-envir_b_816198.html.

Collins, Dwight E., Russell M. Genet, and David Christian. "Crafting a New Narrative to Support Sustainability." In *State of the World 2013: Is Sustainability Still Possible?*, edited by Linda Starke, 218–24. Washington, DC: Island Press, 2013.

Conca, James. "The Largest Clean Coal Power Plant in America Turns to Natural Gas." *Forbes*, July 11, 2017. https://www.forbes.com/sites/james-conca/2017/07/11/the-largest-clean-coal-power-plant-in-america-turns-to-natural-gas/#26c8916f7d26.

Corn, David, and Kate Sheppard. "Obama's Copenhagen Deal: How It Came About—And Why It May Not Be a Real Deal." *Mother Jones*, December 19, 2009. https://www.motherjones.com/environment/2009/12/obamas-copenhagen-deal/.

CQ Magazine. "2019 Appropriations: Interior-Environment." July 30, 2018.

CQ Magazine. "2019 Budget: Environmental Protection Agency." March 12, 2018.

Cushman, John H., Jr. "Talks on Global Warming Treaty Resuming Today." *New York Times*, November 2, 1998. https://www.nytimes.com/1998/11/02/world/talks-on-global-warming-treaty-resuming-today.html.

Daugherty, Owen. "Pelosi Praises Enthusiasm Behind 'Green New Deal' After Seeming to Brush It Off." *The Hill*, February 7, 2019. https://thehill.com/policy/energy-environment/428956-pelosi-praises-enthusiasm-behind-green-new-deal-after-seeming-to-brush-it-off.

Davenport, Coral. "EPA Chief Doubts Consensus View of Climate Change." *New York Times*, March 9, 2017. https://www.nytimes.com/2017/03/09/us/politics/epa-scott-pruitt-global-warming.html.

Davenport, Coral. "No Senate Sequel for Influential Climate Group." *CQ Weekly*, March 29, 2010, 735–36.

Davenport, Coral. "Obama Climate Plan, Now in Court, May Hinge on Error in 1990 Law." *New York Times*, September 25, 2016. https://www.nytimes.com/2016/09/26/us/politics/obama-court-clean-power-plan.html.

Davis, Devra. *When Smoke Ran Like Water: Tales of Environmental Deception and the Battle against Pollution.* Reading, MA: Basic Books, 2002.

Dewey, Scott Hamilton. *Don't Breathe the Air: Air Pollution and U.S. Environmental Politics, 1945–1970.* Environmental History Series, no. 16. College Station: Texas A&M University Press, 2000.

Dimitrov, Radoslav S. "Climate Diplomacy." In *Research Handbook on Climate Governance*, edited by Karin Bäckstrand and Eva Lövbrand, 97–108. Cheltenham, England: Edward Elgar, 2015.

Dimitrov, Radoslav S. "The Paris Agreement on Climate Change: Behind Closed Doors." *Global Environmental Politics* 16, no. 3 (August 2016): 1–11.

Dlugokencky, Ed, and Pieter Tans. "Trends in Atmospheric Carbon Dioxide: Recent Global CO_2." Earth System Research Laboratory, National Oceanic and Atmospheric Administration, viewed July 25, 2017. https://www.esrl.noaa.gov/gmd/ccgg/trends/gl_data.html.

Dodson, John, Xiaoqiang Li, Nan Sun, Pia Atahan, Zhou Xinying, Hanbin Liu, Keliang Zhao, Songmei Hu, and Zemeng Yang. "Use of Coal in the Bronze Age in China." *Holocene* 24, no. 5 (2014): 525–30.

Dunaway, Wilma. "Speculators and Settler Capitalists: Unthinking the Mythology About Appalachian Landholding, 1790–1860." In *Appalachia in the Making: The Mountain South in the Nineteenth Century*, edited by Mary Beth Pudup, Dwight B. Billings, and Altina L. Waller. Chapel Hill: University of North Carolina Press, 1995.

Easterbrook, Steve. "Who First Coined the Term 'Greenhouse Effect'?" *Serendipity* (blog), August 18, 2015. http://www.easterbrook.ca/steve/2015/08/who-first-coined-the-term-greenhouse-effect/.

Eilperin, Juliet, and Josh Dawsey. "Ryan Zinke Resigns as Interior Secretary amid Multiple Investigations." *Chicago Tribune*, December 15, 2018. https://www.chicagotribune.com/news/nationworld/politics/ct-ryan-zinke-resigns-interior-secretary-20181215-story.html.

Ekholm, Nils. "On the Variations of the Climate of the Geological and Historical Past and Their Causes." *Quarterly Journal of the Royal Meteorological Society* 27, no. 117 (1901): 1–62.

Emmanuel, Rahm. *Statement from Mayor Emanuel on Clean Power Plan*. Press release, October 10, Chicago: City of Chicago, Office of the Mayor, 2017. https://www.chicago.gov/city/en/depts/mayor/press_room/press_releases/2017/october/CleanPowerPlan.html.

Environmental Defense Fund. "Overview of EPA Endangerment Finding." Released September 2011. https://www.edf.org/overview-epa-endangerment-finding.

Environmental Protection Agency. "Climate Change Indicators: Atmospheric Concentrations of Greenhouse Gases." Updated April 2016. https://www.epa.gov/climate-indicators/climate-change-indicators-atmospheric-concentrations-greenhouse-gases.

Environmental Protection Agency. "Electric Utility Generating Units: Repealing the Clean Power Plan." https://www.epa.gov/stationary-sources-air-pollution/electric-utility-generating-units-repealing-clean-power-plan.

Environmental Protection Agency. "Endangerment and Cause or Contribute Findings for Greenhouse Gases Under the Section 202(a) of the Clean Air Act." https://www.epa.gov/ghgemissions/endangerment-and-cause-or-contribute-findings-greenhouse-gases-under-section-202a-clean.

Environmental Protection Agency. *EPA's Endangerment Finding: Legal Background*, December 7, 2009. https://www.epa.gov/sites/production/files/2016–08/documents/endangermentfinding_legalbasis.pdf.

Environmental Protection Agency. "Proposal: Affordable Clean Energy (ACE) Rule." August 21, 2018. https://www.epa.gov/stationary-sources-air-pollution/proposal-affordable-clean-energy-ace-rule.

Environmental Protection Agency. "Repeal of Carbon Pollution Emission Guidelines for Existing Stationary Sources: Electric Utility Generating Units:

Proposed Rule." *Federal Register* 82, no. 198 (16 October 2017): 48035–49. https://www.govinfo.gov/content/pkg/FR-2017-12-28/pdf/2017-27793.pdf.

Environmental Protection Agency. "State Guidelines for Greenhouse Gas Emissions from Existing Electric Utility Generating Units: Advance Notice of Proposed Rulemaking." *Federal Register* 82, no. 248 (December 28, 2017): 61507–19. https://www.govinfo.gov/content/pkg/FR-2017-12-28/pdf/2017-27793.pdf.

Eshelman, Robert S. "Cracking Big Coal." *Nation* 290, no. 17 (May 3, 2010): 17–20.

Evelyn, John. *Fumifugium: or, The Inconvenience of the Aer and Smoak of London Dissipated*. Manchester: National Smoke Abatement Society, 1933.

Evelyn, John. *Sylva: or, A Discourse of Forest Trees and the Propagation of Timber*. London: J. Martyn and J. Allestry, printers to the Royal Society, 1670.

"Even Some GOP Lawmakers Don't Like Trump's Cuts to Environmental Programs." *CQ Magazine*, June 12, 2017..

Feely, Richard A., Christopher L. Sabine, Kitack Lee, Will Berelson, Joanie Kleypas, Victoria J. Fabry, and Frank J. Millero. "Impact of Anthropogenic CO_2 on the $CaCO_2$ System in the Oceans." *Science* 305, no. 5682 (July 15, 2004): 362–66.

Flynn, Lucy. "Boris Johnson's Decision Not to Divest Goes against His Climate Goals for London." *Guardian*, May 21, 2015. https://www.theguardian.com/environment/2015/may/21/boris-johnsons-decision-not-to-divest-goes-against-his-climate-goals-for-london.

"For the Record: How the House Voted June 26." *CQ Weekly*, July 6, 2009, 1588–92.

Fountain, Henry. "Corralling Carbon before It Belches from Stack." *New York Times*, July 21, 2014. https://www.nytimes.com/2014/07/22/science/corralling-carbon-before-it-belches-from-stack.html.

Fountain, Henry. "Turning Carbon Dioxide Into Rock, and Burying It." *New York Times*, February 9, 2015. https://www.nytimes.com/2015/02/10/science/burying-a-mountain-of-co2.html.

Fried, Lauha, ed. *Global Wind Statistics 2017*. Brussels: Global Wind Energy Council, 2018.

Fried, Lauha, Shruti Shukla, Steve Sawyer, and Sven Teske, eds. *Global Wind Energy Outlook 2016*. Brussels: Global Wind Energy Council, 2016.

Friedman, Lisa. "Andrew Wheeler, New EPA Chief, Details His Energy Lobbying Past." *New York Times*, August 1, 2018. https://www.nytimes.com/2018/08/01/climate/andrew-wheeler-epa-lobbying.html.

Friedman, Lisa. "Former Trump Aide Calls Paris Climate Accord 'A Good Republican Agreement.'" *New York Times*, February 22, 2018. https://www.nytimes.com/2018/02/22/climate/george-david-banks.html.

Friedman, Lisa. "How a Coal Baron's Wish List Became President Trump's To-Do List." *New York Times*, January 9, 2018. https://www.nytimes.com/2018/01/09/climate/coal-murray-trump-memo.html.

Friedman, Lisa. "Trump Says He'll Nominate Andrew Wheeler to Head the EPA." *New York Times*, November 16, 2018. https://www.nytimes.com/2018/11/16/climate/trump-andrew-wheeler-epa.html.

Friedman, Lisa, and Brad Plumer. "EPA Announces Repeal of Major Obama-Era Carbon Emissions Rule." *New York Times*, October 9, 2017. https://www.nytimes.com/2017/10/09/climate/clean-power-plan.html.

Geuss, Megan. "$7.5 Billion Kemper Power Plant Suspends Coal Gasification." *Ars Technica*, June 28, 2017. https://arstechnica.com/information-technology/2017/06/7–5-billion-kemper-power-plant-suspends-coal-gasification/.

Glassman, James K. "Administration in the Balance." *Wall Street Journal*, Eastern edition, March 8, 2001.

Global CCS Institute. *The Global Status of CCS: 2017*. Melbourne: Global CCS Institute, 2017.

Global CCS Institute. *Projects Database: Large-Scale CCS Facilities*. Melbourne: Global CCS Institute, 2018. http://www.globalccsinstitute.com/projects/large-scale-ccs-projects.

Global Covenant of Mayors for Climate & Energy. "Global Covenant of Mayors for Climate & Energy's Cities Have Committed to Reducing Emissions Equal to Removing All U.S. Cars Off the Road." Press release, September 10, 2018. https://www.globalcovenantofmayors.org/wp-content/uploads/2018/09/GCoM-Data-Press-Release_GCAS_.pdf.

Goldenberg, Suzanne. "Barack Obama's U.S. Climate Change Bill Passes Key Congress Vote." *Guardian*, June 26, 2009. https://www.theguardian.com/environment/2009/jun/27/barack-obama-climate-change-bill.

Goldenberg, Suzanne, and Helena Bengtsson. "Biggest US Coal Company Funded Dozens of Groups Questioning Climate Change." *Guardian*, June 13, 2016. https://www.theguardian.com/environment/2016/jun/13/peabody-energy-coal-mining-climate-change-denial-funding.

Gottlieb, Robert. *Forcing the Spring: The Transformation of the American Environmental Movement*. Rev. ed. Washington, DC: Island Press, 2005.

Grandia, Kevin. "Leaked Clean Coal Strategy Memo to Peabody Energy." *DeSmog* (blog), January 16, 2009. https://www.desmogblog.com/leaked-clean-coal-strategy-memo-peabody-energy.

Grandoni, Dino, Dennis Brady, and Chris Mooney. "EPA's Scott Pruitt Asks Whether Global Warming 'Necessarily Is a Bad Thing.'" *Washington Post*, February 8, 2018. https://www.washingtonpost.com/news/energy-environment/wp/2018/02/07/scott-pruitt-asks-if-global-warming-necessarily-is-a-bad-thing.

Gyorgy, Anna, and friends. *No Nukes: Everyone's Guide to Nuclear Power*. Boston: South End Press, 1979.

Haagen-Smit, Arie. "Formation of Ozone in Los Angeles Smog." In *Proceedings of the Second National Air Pollution Symposium*, 54–56. Pasadena, May 5–6, 1952.

Hakim, Danny. "States Plan Suit to Prod U.S. on Global Warming." *New York Times*, October 3, 2003. https://www.nytimes.com/2003/10/04/business/states-plan-suit-to-prod-us-on-global-warming.html.

Hansen, James. *Storms of My Grandchildren: The Truth About the Coming Climate Catastrophe and Our Last Chance to Save Humanity*. Illustrations by Makiko Sato. New York: Bloomsbury USA, 2009.

Hansen, James, Reto Ruedy, Makiko Sato, and Kenneth Lo. "Global Surface Temperature Change." *Reviews of Geophysics* 48, no. 4 (2010), https://agupubs.onlinelibrary.wiley.com/doi/epdf/10.1029/2010RG000345.

Hansen, James, Makiko Sato, Pushker Kharecha, David Beerling, Robert Berner, Valerie Masson-Delmotte, Mark Pagani, Maureen Raymo, Dana L. Royer, and James C. Zachos. "Target Atmospheric CO_2: Where Should Humanity Aim?" *Open Atmospheric Science Journal* 2, no. 1 (December 2008): 217–31.

Hardin, Garrett. "The Tragedy of the Commons." *Science* 162 (1968): 1243–48.

Hartmann, Dennis L., and Marc L. Michelsen. "No Evidence for Iris." *Bulletin of the American Meteorological Society* 83, no. 2 (February 2002): 249–54.

Haszeldine, R. Stuart. "Carbon Capture and Storage: How Green Can Black Be?" *Science* 325, no. 5948 (September 25, 2009): 1647–52.

Heinberg, Richard. *Blackout: Coal, Climate and the Last Energy Crisis*. Gabriola Island, BC: New Society, 2009.

Henry, Devin. "House Bill Would Cut EPA Funding by $528M." *The Hill*, July 11, 2017. https://thehill.com/policy/energy-environment/341507-house-bill-would-cut-epa-funding-by-528m.

Hertsgaard, Mark. "Emissions Impossible." *Mother Jones* 37, no. 3 (May-June 2012): 36–46.

"Highlights of the House Climate Bill." *CQ Weekly*, June 29, 2009, 1517.

Hockett, Robert. "The Green New Deal Is What Our Planet Has Been Waiting For." *The Hill*, February 7, 2019.

Hofacker, Carl. "Interior Secretary Ryan Zinke Accuses Democrat Who Wants Him to Resign of 'Drunken and Hostile Behavior.'" *USA Today*, November 30, 2018. https://www.usatoday.com/story/news/politics/2018/11/30/ryan-zinke-accuses-rep-paul-grijalva-drunken-hostile-behavior/2163291002/.

Horan, James D. *The Pinkertons: The Detective Dynasty That Made History*. New York: Crown, 1968.

Hossain, Farhana, Amanda Cox, John McGrath, and Stephan Weitberg. "The Stimulus Plan: How to Spend $787 Billion." *New York Times*, February 17, 2009. https://www.nytimes.com/interactive/projects/44th_president/stimulus#.

Houghton, J. T., Y. Ding, D. J. Griggs, M. Noguer, P. J. van der Linden, X. Dai, K. Maskell, and C. A. Johnson, eds. *Climate Change 2001: The Scientific Basis, Contribution of Working Group I to the Third Assessment Report of the Intergovernmental Panel on Climate Change*. New York: Cambridge University Press, 2001.

Hoyos, C. D., P. A. Agudelo, P. J. Webster, and J. A. Curry. "Deconvolution of the Factors Contributing to the Increase in Global Hurricane Intensity." *Science* 312, no. 5770 (April 7, 2006): 94–97.

Interdepartmental Committee on Air Pollution. *Air Pollution: Proceedings of the United States Technical Conference on Air Pollution.* Sponsored by the Interdepartmental Committee on Air Pollution. New York: McGraw-Hill, 1952.

Intergovernmental Panel on Climate Change. "Policymaker Summary of Working Group I (Scientific Assessment of Climate Change)." In *Climate Change: The IPCC 1990 and 1992 Assessments, First Assessment Report Overview and Policymaker Summaries and 1992 IPCC Supplement.*Geneva: Intergovernmental Panel on Climate Change, 1992. https://www.ipcc.ch/report/climate-change-the-ipcc-1990-and-1992-assessments/.

Intergovernmental Panel on Climate Change. *Special Report: Global Warming of 1.5°C,* 2018. https://www.ipcc.ch/sr15/.

International Energy Agency. *Renewable Energy Essentials: Hydropower.* November 8, 2010. http://www.iea.org/publications/freepublications/publication/hydropower_essentials.pdf.

International Energy Agency. *World Energy Outlook 2017.* Paris: International Energy Agency, 2017. http://www.iea.org/weo2017/.International Energy Association. *Carbon Capture and Storage Roadmap,* n.d. https://www.iea.org/publications/freepublications/publication/CCS_roadmap_foldout.pdf.

International Monetary Fund. "World Economic and Financial Surveys: World Economic Database." https://www.imf.org/external/pubs/ft/weo/2017/02/weodata/index.aspx.

Irby-Massie, Georgia L., and Paul T. Keyser. *Greek Science of the Hellenistic Era: A Sourcebook.* London: Routledge, 2002.

Irfan, Umair. "Interior Secretary Ryan Zinke Might Face a Criminal Investigation."*Vox,*November 5, 2018. https://www.vox.com/2018/10/31/18044860/ryan-zinke-interior-investigation-ethics-justice.

Itaipu Binacional. "Energy." https://www.itaipu.gov.br/en/energy/energy.

Jacobson, Louis. "How Much Have Republicans Cut EPA Budget, Staff in Two Years?" *Politifact,* April 26, 2018. https://www.politifact.com/truth-o-meter/statements/2018/apr/26/evan-jenkins/how-much-have-republicans-cut-epa-budget-staff-two/.

Jalonick, Mary Clare. "Defeat of Senate Global Warming Bill Highlights Worries over Economic Impact." *CQ Weekly,* November 1, 2003, 2709–10.

"James Hansen: Taking Heat for Decades." *Bulletin of the Atomic Scientists* 69, no. 4 (July 2013): 1–8.

Jevons, William Stanley. *The Coal Question: An Enquiry Concerning the Progress of the Nation, and the Probable Exhaustion of Our Coal-Mines.* London: Macmillan, 1866.

Jones, Brad. "New Multi-Industry Coalition Aligns to Advocate Energy Security and Environmental Stewardship." Press release, American Coalition for Clean Coal Electricity, April 17, 2008. https://www.businesswire.com/

news/home/20080417006084/en/New-Multi-Industry-Coalition-Aligns
-Advocate-Energy-Security.

Jones, Charles O. *Clean Air: The Policies and Politics of Pollution Control*. Pittsburgh:
University of Pittsburgh Press, 1975.

Jordans, Frank. "Talks Adopt 'Rulebook' to Put Paris Climate Deal into Action."
New York Times, December 16, 2018. https://www.apnews.com/92da6d1
65f6c4addbbbb82ac7a3eed84.

Kallanish Energy. "Trump's EPA Releases Affordable Clean Energy Rule." August
22, 2018. http://www.kallanishenergy.com/2018/08/22/trumps-epa-releases
-affordable-clean-energy-rule/.

Keeling, Charles D. "The Concentration and Isotopic Abundances of Carbon
Dioxide in the Atmosphere." *Tellus* 12, no. 2 (1960): 200–3.

Kelly, Sharon. "Bosses at World's Most Ambitious Clean Coal Plant Kept Problems
Secret for Years." *Guardian*, March 2, 2018. https://www.theguardian.com/
us-news/2018/mar/02/clean-coal-kemper-plant-mississippi-problems.

Kelly, Sharon. "How America's Clean Coal Dream Unravelled." *Guardian*, March
2, 2018. https://www.theguardian.com/environment/2018/mar/02/clean-coal
-america-kemper-power-plant.

Kerry, John. *Keystone XL Pipeline Permit Determination*. Washington, DC: Department of State, 2015. https://2009-2017.state.gov/secretary/remarks/2015/11/
249249.htm.

"Keystone XL Pipeline Overview." *Congressional Digest* 90, no. 10 (December
2011): 290–95.

King, Hobart. "History of Energy Use in the United States." Geology.com. http://
geology.com/articles/history-of-energy-use/.

Klare, Michael. "Bush-Cheney Energy Strategy: Procuring the Rest of the World's
Oil." *Foreign Policy in Focus*, January 2004. https://www.thegoldenaura
.com/wp-content/uploads/2004/09/bushcheneystrategy.pdf.

Konisky, David M., and Neal D. Woods. "Environmental Policy, Federalism, and
the Obama Presidency." *Publius: The Journal of Federalism* 46, no. 3 (Summer 2016): 366–91.

Kopp, Otto C. "Coal." In *Encyclopedia Britannica Online*, 2015. http://www
.britannica.com/science/coal-fossil-fuel/Resources-and-reserves.

Kreis, Steven. "Lecture 17: The Origins of the Industrial Revolution in England."
The History Guide. Last updated April 24, 2017. http://www.historyguide
.org/intellect/lecture17a.html.

Krieg, Gregory. "Who Is Alexandria Ocasio-Cortez?" *CNN Politics*, June 27, 2018.
https://www.cnn.com/2018/06/27/politics/who-is-alexandria-ocasio
-cortez/index.html.

Krier, James E., and Edmund Ursin. *Pollution and Policy: A Case Essay on California and Federal Experience with Motor Vehicle Air Pollution, 1940–1975*.
Berkeley: University Of California Press, 1977.

Lade, Gabriel E., C-Y Cynthia Lin Lawell, and Aaron Smith. "Designing Climate
Policy: Lessons from the Renewable Fuel Standard and the Blend Wall."

American Journal of Agricultural Economics 100, no. 2 (March 2018): 585–99.

Lampitt, R. S., E. P. Achterberg, T. R. Anderson, J. A. Hughes, M. D. Iglesias-Rodriguez, B. A. Kelly-Gerreyn, M. Lucas, E. E. Popova, R. Sanders, J. G. Shepherd, D. Smythe-Wright, and A. Yool. "Ocean Fertilization: A Potential Means of Geoengineering?" *Philosophical Transactions of the Royal Society A: Mathematical, Physical and Engineering Sciences* 366, no. 1882 (2008): 3919–45.

Latham, John, Philip Rasch, Chih-Chieh (Jack) Chen, Laura Kettles, Alan Gadian, Andrew Gettelman, Hugh Morrison, Keith Bower, and Tom Choularton. "Global Temperature Stabilization via Controlled Albedo Enhancement of Low-Level Maritime Clouds." *Philosophical Transactions of the Royal Society A: Mathematical, Physical and Engineering Sciences* 366, no. 1882 (2008): 3969–87.

Lavelle, Marianne. "The 'Clean Coal' Lobbying Blitz: As Climate Change Hearings Begin on Capitol Hill, a Coal Industry Group Flexes New-Found Muscle." *Center for Public Integrity*, April 21, 2009. https://www.public integrity.org/2009/04/21/2885/clean-coal-lobbying-blitz.

Layden, Logan. "Court Losses Won't Deter Attorney General Scott Pruitt in His Fight with the EPA." *StateImpact Oklahoma*, June 12, 2014. https://stateim-pact.npr.org/oklahoma/2014/06/12/court-losses-wont-deter-attorney -general-scott-pruitt-in-his-fight-with-the-epa/.

Lee, Kai N., William R. Freudenburg, and Richard B. Howarth. *Humans in the Landscape: An Introduction to Environmental Studies.* New York: W. W. Norton, 2013.

Levy, Robert A. "Who Elected Lisa Jackson?" *CATO Policy Report* 33, no. 2 (March-April 2011): 2.

Lewis, Mario, Grover Norquist, Matt Kibbe, Duane Parde, Larry Hart, Eric Heidenreich, Michael Bowman, et al. Public comment, November 24, 2008. "Regulating Greenhouse Gases under the Clean Air Act," No. EPA-HQ-OAR-2008–0318. https://www.heartland.org/_template-assets/doc uments/publications/24270.pdf.

Li, Kaifeng, Cheng Zhu, Li Wu, and Linyan Huang. "Problems Caused by the Three Gorges Dam Construction in the Yangtze River Basin: A Review." *Environmental Reviews* 21, no. 3 (September 2013): 127–35.

Lichtenstein, Nelson. "Two Roads Forward for Labor: The AFL-CIO's New Agenda." *Dissent* 61, no. 1 (2014): 54–58.

Likens, Gene E., and F. Herbert Bormann. "Acid Rain: A Serious Regional Environmental Problem." *Science* 184, no. 4142 (June 14, 1974): 1176–79.

Likens, Gene E., F. Herbert Bormann, and Noye M. Johnson. "Acid Rain." *Environment: Science and Policy for Sustainable Development* 14, no. 2 (March 1, 1972): 33–40.

Lin, Bing, Bruce A. Wielicki, Lin H. Chambers, Yongxiang Hu, and Kuan-Man Xu. "The Iris Hypothesis: A Negative or Positive Cloud Feedback?" *Journal of Climate* 15, no. 1 (January 2002): 3–7.

Lindsey, Rebecca, and LuAnn Dahlman. "Climate Change: Global Temperature," updated August 1, 2018. https://www.climate.gov/news-features/under standing-climate/climate-change-global-temperature.

Lindzen, Richard S. "Can Increasing Carbon Dioxide Cause Climate Change?" *PNAS: Proceedings of the National Academy of Sciences of the United States of America* 94, no. 16 (August 5, 1997): 8335–42. http://www.pnas.org/content/94/16/8335.

Lindzen, Richard S. "Climate Physics, Feedbacks, and Reductionism (and When Does Reductionism Go Too Far?)." *European Physical Journal Plus* 127, no. 5 (2012): Article 52.

Lindzen, Richard S. "Some Coolness Concerning Global Warming." *Bulletin of the American Meteorological Society* 71, no. 3 (March 1, 1990): 288–99.

Lindzen, Richard S., Ming-Dah Chou, and Arthur Y. Hou. "Comment on 'No Evidence for Iris.'" *Bulletin of the American Meteorological Society* 83, no. 9 (September 2002): 1345–49.

Lindzen, Richard S., Ming-Dah Chou, and Arthur Y. Hou. "Does the Earth Have an Adaptive Infrared Iris?" *Bulletin of the American Meteorological Society* 82, no. 3 (March 2001): 417.

Liptak, Adam, and Coral Davenport. "Supreme Court Deals Blow to Obama's Efforts to Regulate Coal Emissions." *New York Times*, February 9, 2016. https://www.nytimes.com/2016/02/10/us/politics/supreme-court-blocks -obama-epa-coal-emissions-regulations.html.

Locke, John. "The Second Treatise." In *Two Treatises of Government*, edited by Peter Laslett. Reprint, New York: Mentor Books, 1965.

London, Jonathan, Alex Karner, Julie Sze, Dana Rowan, Gerardo Gambirazzio, and Deb Niemeier. "Racing Climate Change: Collaboration and Conflict in California's Global Climate Change Policy Arena." *Global Environmental Change* 23, no. 4 (2013): 791–99.

Lovelock, James. "Nuclear Power Is the Only Green Solution." *Independent*, May 23, 2004. https://www.independent.co.uk/voices/commentators/james -lovelock-nuclear-power-is-the-only-green-solution-564446.html.

Lovins, Amory B., and John H. Price. *Non-Nuclear Futures: The Case for an Ethical Energy Strategy*. Friends of the Earth Energy Papers. San Francisco: Friends of the Earth International, 1975.

Lowe, Jeanne R. *Cities in a Race with Time: Progress and Poverty in America's Renewing Cities*. New York: Random House, 1967.

Lukas, J. Anthony. *Big Trouble: A Murder in a Small Western Town Sets Off a Struggle for the Soul of America*. New York: Simon & Schuster, 1997.

Lutey, Tom. "Zinke Resigns Delegate Post over Public Lands Disagreement; Still Will Speak at RNC." *Billings Gazette*, July 15, 2016. https://billings gazette.com/news/local/zinke-resigns-delegate-post-over-public-lands -disagreement-still-will/article_8109f084-d199-50dd-b223-9fd3557a738d .html.

Main, Douglas. "EPA, Interior Plan to Cut More Than 5,000 Staff by 2018." *Newsweek*, June 21, 2017. https://www.newsweek.com/epa-interior-plan-cut-more-5000-staff-2018-627998.

Mallard, Megan Suzanne. "Atlantic Hurricanes and Climate Change." PhD diss., North Carolina State University, 2011. http://www.lib.ncsu.edu/resolver/1840.16/7184.

Malone, Patrick M. *Waterpower in Lowell: Engineering and Industry in Nineteenth-Century America*. Johns Hopkins Introductory Studies in the History of Technology. Baltimore: Johns Hopkins University Press, 2009.

Mann, Thomas E., and Norman J. Ornstein. *It's Even Worse Than It Looks: How the American Constitutional System Collided with the New Politics of Extremism*. New York: Basic Books, 2013.

Marsh, George P. *Man and Nature, or Physical Geography as Modified by Human Action*. New York: Charles Scribner, 1864.

Marshall, Robert. "The Wilderness as a Minority Right." *U.S. Forest Service Bulletin*, August 27, 1928, 5–6.

Martelle, Scott. *Blood Passion: The Ludlow Massacre and Class War in the American West*. New Brunswick, NJ: Rutgers University Press, 2007.

Mayhew, David R. *Divided We Govern: Party Control, Lawmaking, and Investigations, 1946–2002*. New Haven, CT: Yale University Press, 2005.

McGovern, George S., and Leonard F. Guttridge. *The Great Coalfield War*. Maps by Samuel H. Bryant. Boston: Houghton Mifflin, 1972.

McKibben, Bill. "Copenhagen: Things Fall Apart and an Uncertain Future Looms." *Yale Environment 360*, December 21, 2009. http://e360.yale.edu/feature/copenhagen_things_fall_apart_and_an_uncertain_future_looms/2225/.

McKibben, Bill. *The End of Nature*. New York: Random House, 1989.

McKibben, Bill. "The Reckoning." *Rolling Stone* no. 1162 (August 2, 2012): 52–60.

McNeill, Brian. "Forged Letters Scandal Widens." *Daily Progress*, August 5, 2009. http://www.dailyprogress.com/news/forged-letters-scandal-widens/article_7dfdb333-4860-57b2-89fd-4852ac6425fe.html.

Meyer, Balthasar Henry, Caroline Elizabeth MacGill, et al. *History of Transportation in the United States before 1860*. Prepared under the direction of Balthasar Henry Meyer by Caroline E. MacGill and a staff of collaborators. Contributions to American Economic History [3]. Washington, DC: Carnegie Institution of Washington, 1917; reprinted New York: Peter Smith, 1948.

Miller, Stephen L. "Coal Industry Strategy Letter to CEO of Peabody Energy." June 18, 2004. https://www.desmogblog.com/sites/beta.desmogblog.com/files/Coal%20Industry%20Strategy%20Letter%20To%20CEO%20of%20Peabody%20Energy.pdf.

Mintzer, Irving M., and J. Amber Leonard, eds. *Negotiating Climate Change: The Inside Story of the Rio Convention*. Cambridge Studies in Energy and Environment. Cambridge, UK: Cambridge University Press, 1994.

Monbiot, George. "Why Fukushima Made Me Stop Worrying and Love Nuclear Power." *Guardian*, March 21, 2011. https://www.theguardian.com/com mentisfree/2011/mar/21/pro-nuclear-japan-fukushima.

Monies, Paul. "Oklahoma Attorney General, 11 Others File Lawsuit against EPA Over 'Sue and Settle' Tactics." *NewsOK*, July 17, 2013. https://newsok .com/article/3862959/oklahoma-attorney-general-11-others-file-lawsuit -against-epa-over-sue-and-settle-tactics.

Montrie, Chad. *To Save the Land and People: A History of Opposition to Surface Coal Mining in Appalachia*. Chapel Hill: University of North Carolina Press, 2003.

Morn, Frank. *"The Eye That Never Sleeps": A History of the Pinkerton National Detective Agency*. Bloomington: Indiana University Press, 1982.

Morton, Oliver. "A New Day Dawning?: Silicon Valley Sunrise." *Nature* 443 (September 6, 2006): 19.

Muir, John. *Let Everyone Help to Save the Famous Hetch-Hetchy Valley and Stop the Commercial Destruction Which Threatens Our National Parks: A Brief Statement of the Hetch-Hetchy Case to Date.*. n.p., 1909. http://lcweb2.loc.gov/gc/ amrvg/vg50/vg50.html.

Muir, John. *Meditations of John Muir: Nature's Temple*. Compiled and edited by Chris Highland. Berkeley: Wilderness Press, 2001.

Mulkern, Anne C. "'Citizen Army' Carries Coal's Climate Message to Hinterlands." *New York Times*, August 6, 2009. http://www.nytimes.com/gwire/ 2009/08/06/06greenwire-citizen-army-carries-coals-climate-message -to-39075.html.

Murphy, Tim. "Trump's Interior Nominee Was for Climate Action before He Was against It." *Mother Jones*, December 14, 2016. https://www.motherjones .com/politics/2016/12/ryan-zinke-donald-trump-climate-change/.

National Council for Science and the Environment. "Building Climate Solutions." 14th National Conference and Global Forum on Science, Policy and the Environment, January 28–31, 2014, Washington, DC. https:// ncseconference.org/wp-content/uploads/2017/06/2014-conference -program.pdf.

National Energy Policy Development Group. *Reliable, Affordable, and Environmentally Sound Energy for America's Future*. Washington, DC: White House, 2001. http://www.dtic.mil/dtic/tr/fulltext/u2/a392171.pdf.

Nawaguna, Elvina. "Democrats Unveil Green New Deal That Would Push Government to Make Radical Changes." *Roll Call*, February 7, 2019. https:// www.rollcall.com/news/congress/democrats-offer-green-new-deal-resolu tion-for-economic-overhaul.

Niemeier, Ulrike, and Simone Tilmes. "Sulfur Injections for a Cooler Planet." *Science* 357, no. 6348 (July 21, 2017): 246–48.

NOAA National Centers for Environmental Information. "Climate at a Glance: Global Time Series." 2017. http://www.ncdc.noaa.gov/cag/global/time -series.

Nordhaus, Ted, and Michael Shellenberger. *Break Through: From the Death of Environmentalism to the Politics of Possibility.* Boston: Houghton Mifflin, 2007.

Obama, Barack. "A New Chapter on Climate Change." Video address, 2008. https://www.youtube.com/watch?v=hvG2XptIEJk.

Obama, Barack. *Presidential Memorandum: A Comprehensive Federal Strategy on Carbon Capture and Storage.* Washington, DC: White House, Office of the Press Secretary, 2010. https://obamawhitehouse.archives.gov/the-press-office/presidential-memorandum-a-comprehensive-federal-strategy-carbon-capture-and-storage.

Obama, Barack. *Remarks by the President in Announcing the Clean Power Plan.* Washington, DC: White House, Office of the Press Secretary, 2015. https://obamawhitehouse.archives.gov/the-press-office/2015/08/03/remarks-president-announcing-clean-power-plan.

Obama, Barack. *Remarks by the President in the State of the Union Address.* Washington, DC: White House, Office of the Press Secretary, 2013. https://obamawhitehouse.archives.gov/the-press-office/2013/02/12/remarks-president-state-union-address.

Obama, Barack. *Remarks of President Barack Obama in the State of the Union Address—as Prepared for Delivery.* Washington, DC: White House, Office of the Press Secretary, 2011. http://www.whitehouse.gov/the-press-office/2011/01/25/remarks-president-barack-obama-state-union-address-prepared-delivery.

O'Leary, Rosemary. "Environmental Policy in the Courts." In *Environmental Policy: New Directions for the Twenty-First Century.* 10th ed., edited by Norman J. Vig and Michael E. Kraft, 144–67. Thousand Oaks, CA: CQ Press, 2019.

Olson, Mancur, Jr. *The Logic of Collective Action: Public Goods and the Theory of Groups.* 2nd ed. Cambridge, MA: Harvard University Press, 1971.

Ostrom, Elinor. *Governing the Commons: The Evolution of Institutions for Collective Action.* Cambridge, UK: Cambridge University Press, 1990.

Parker, Charles F., and Christer Karlsson. "The UN Climate Change Negotiations and the Role of the United States: Assessing American Leadership from Copenhagen to Paris." *Environmental Politics* 27, no. 3 (February 22, 2018): 1–22.

Parson, Edward A. "Assessing UNCED and the State of Sustainable Development." *Proceedings of the American Society of International Law Annual Meeting* 87 (1993): 508–13.

Parzen, Julia. *Lessons Learned: Creating the Chicago Climate Action Plan.* Chicago: Chicago Climate Action Plan, 2009. http://www.chicagoclimateaction.org/filebin/pdf/LessonsLearned.pdf.

Payne, Daniel G., and Richard S. Newman, eds. *The Palgrave Environmental Reader.* New York: Palgrave Macmillan, 2005.

Pibel, Doug, and Van Jones. "Van Jones: Why I'm Going to Washington." *Yes!* March 10, 2009. https://www.yesmagazine.org/issues/food-for-everyone/van-jones-why-i2019m-going-to-washington.

Pike, Robert E. *Tall Trees, Tough Men*. New York: W. W. Norton, 1999.

Pinchot, Gifford. *The Fight for Conservation*. London: Hodder & Stoughton, 1910.

Plumer, Brad. "Climate Negotiators Reach an Overtime Deal to Keep Paris Pact Alive." New York Times, December 15, 2018. https://www.nytimes .com/2018/12/15/climate/cop24-katowice-climate-summit.html.

PMEL Carbon Program. National Oceanic and Atmospheric Administration. "Ocean Acidification: The Other Carbon Dioxide Problem." https://pmel .noaa.gov/co2/story/Ocean+Acidification.

PMEL Carbon Program. National Oceanic and Atmospheric Administration. "What is Ocean Acidification?" PMEL Carbon Program. https://pmel .noaa.gov/co2/story/What+is+Ocean+Acidification%3F.

Potsdam Institute for Climate Impact Research. "Ambitions of Only Two Developed Countries Sufficiently Stringent for 2°C." February 3, 2010. https://www.pik-potsdam.de/news/in-short/archive/2010/ambition -of-only-two-developed-countries-sufficiently-stringent-for-2b0c.

Poynting, J. H. "On Prof. Lowell's Method for Evaluating the Surface-Temperatures of the Planets; with an Attempt to Represent the Effect of Day and Night on the Temperature of the Earth." *Philosophical Magazine* 14, no. 84 (December 1907): 749–60.

Price, Trevor J. "James Blyth: Britain's First Modern Wind Power Pioneer." *Wind Engineering* 29, no. 3 (May 1, 2005): 191–200.

"Promises about Environment on the Obameter." *St. Petersburg Journal: Politifact*. http://www.politifact.com/truth-o-meter/promises/obameter/subjects/ environment/.

Rajamani, Lavanya. "Ambition and Differentiation in the 2015 Paris Agreement: Interpretive Possibilities and Underlying Politics." *International & Comparative Law Quarterly* 65, no. 2 (April 2016): 493–514.

Raju, Manu. "Anxious Coal Industry Deep in Emissions Debate." *CQ Weekly*, March 19, 2007, 795–96.

Räthzel, Nora, and David Uzzell. "Trade Unions and Climate Change: The Jobs versus Environment Dilemma." *Global Environmental Change* 21, no. 4 (October 2011): 1215–23.

Raymond, Leigh. *Reclaiming the Atmospheric Commons: The Regional Greenhouse Gas Initiative and a New Model of Emissions Trading*. Cambridge, MA: MIT Press, 2016.

Reilly, William K. "Reflections on U.S. Environmental Policy: An Interview with William K. Reilly." Produced by Dan Fiorino and Gordon Binder. http:// www.epaalumni.org/userdata/pdf/3E0FC143699CD31B.pdf.

REN21. *Renewables 2016: Global Status Report*. Paris: REN Secretariat, 2016. http://www.ren21.net/wp-content/uploads/2016/06/GSR_2016_Full _Report_REN21.pdf.

Revelle, Roger, and Hans E. Suess. "Carbon Dioxide Exchange between Atmosphere and Ocean and the Question of an Increase of Atmospheric CO_2 during the Past Decades." *Tellus* 9, no. 1 (1957): 18–27.

Reynolds, Terry S. *Stronger than a Hundred Men: A History of the Vertical Water Wheel*. Johns Hopkins Studies in the History of Technology, no. 7. Baltimore: Johns Hopkins University Press, 1983.

RGGI advocates. "Stronger RGGI for a Clean Energy Economy: What is RGGI." https://www.cleanenergyeconomy.us/#about-rggi.

Rhodium Group Energy and Climate Staff. "Preliminary U.S. Emissions Estimates for 2018." January 8, 2019. https://rhg.com/research/preliminary-us-emissions-estimates-for-2018/.

Richmond, Glenn. "The Last Logjam: Machias River Log Drive." *Down East: The Magazine of Maine*, May 1971. https://downeast.com/archives-may-1971/.

Ripley, Randall B. "Congress and Clean Air: The Issue of Enforcement, 1963." In *Congress and Urban Problems: A Casebook on the Legislative Process*, edited by Frederic N. Cleaveland, 224–78. Washington, DC: Brookings Institution, 1969.

Roberts, David. "Turns Out the World's First 'Clean Coal' Plant Is a Backdoor Subsidy to Oil Producers." *Grist*, March 31, 2015. https://grist.org/climate-energy/turns-out-the-worlds-first-clean-coal-plant-is-a-backdoor-subsidy-to-oil-producers/.

Robock, Alan, Luke Oman, and Georgiy L. Stenchikov. "Regional Climate Responses to Geoengineering with Tropical and Arctic SO_2 Injections." *Journal of Geophysical Research: Atmospheres* 113, no. D16 (August 2008). doi:10.1029/2008JD010050

Rochelle, Gary T. "Amine Scrubbing for CO_2 Capture." *Science* 325, no. 5948 (September 25, 2009): 1652–54.

Rosenbaum, Walter A. *Environmental Politics and Policy*. 9th ed. Washington, DC: CQ Press, 2014.

Rosenbaum, Walter A. *Environmental Politics and Policy*. 10th ed. Washington, DC: CQ Press, 2017.

Rosencranz, Armin. "U.S. Climate Change Policy under G. W. Bush." *Golden Gate University Law Review* 32, no. 4 (January 2002): 479–91.

Ross, Benjamin, and Steven Amter. *The Polluters: The Making of Our Chemically Altered Environment*. Oxford: Oxford University Press, 2010.

Rudalevige, Andrew. "The Contemporary Presidency: The Obama Administrative Presidency: Some Late-Term Patterns." *Presidential Studies Quarterly* 46, no. 4 (December 2016): 868–90.

Saad, Lydia. "Global Warming Concern at Three-Decade High in U.S." *Gallup*, March 14, 2017. http://www.gallup.com/poll/206030/global-warming-concern-three-decade-high.aspx.

Sanger, Clyde. "Environment and Development." *International Journal* 28, no. 1 (1972): 103–20.

Schatz, Thomas A. Public comment. "Regulating Greenhouse Gases under the Clean Air Act." No. EPA-HQ-OAR-2008–0318, n.d. https://www.heartland.org/_template-assets/documents/publications/24254.pdf.

Schissler, Andrew. "Strip Mining." *Encyclopedia of Earth*, 2006. http://www
.eoearth.org/view/article/156280/.

Schwab, Jim. *Deeper Shades of Green: The Rise of Blue-Collar and Minority Environmentalism in America*. San Francisco: Sierra Club Books, 1994.

"Scott Pruitt: Controversial Trump Environment Nominee Sworn In." *BBC News*,
February 17, 2017. https://www.bbc.com/news/world-us-canada-39010374.

Sengupta, Somini, Melissa Eddy, Chris Buckley, and Alissa J. Rubin. "As Trump
Exits Paris Agreement, Other Nations Are Defiant." *New York Times*, June
1, 2017. https://www.nytimes.com/2017/06/01/world/europe/climate-paris
-agreement-trump-china.html.

Seyfang, Gill. "Environmental Mega-Conferences: From Stockholm to Johannesburg and Beyond." *Global Environmental Change* 13, no. 3 (2003):
223–28.

Shapecoff, Philip. "Global Warming Has Begun, Expert Tells Senate." *New York
Times*, June 24, 1988. http://www.nytimes.com/1988/06/24/us/global
-warming-has-begun-expert-tells-senate.html.

Shear, Michael D. "Trump Will Withdraw U.S. from Paris Climate Agreement."
New York Times, June 1, 2017.

Shear, Michael D., and Alison Smale. "Leaders Lament U.S. Withdrawal, but Say It
Won't Stop Climate Efforts." *New York Times*, June 2, 2017. https://www
.nytimes.com/2017/06/02/climate/paris-climate-agreement-trump.html.

Sierra Club. "About Us: Coal is an Outdated, Backward, and Dirty 19th-Century
Technology." Beyond Coal. https://content.sierraclub.org/coal/about-the
-campaign.

Sierra Club. "2013: A Landmark Year for Clean Energy; Twilight for Coal." *Compass: Pointing the Way to a Clean Energy Future*, December 12, 2013. http://
sierraclub.typepad.com/compass/2013/12/2013-a-landmark-year
-for-clean-energy-twilight-for-coal-.html.

Sierra Club. "Victories." Beyond Coal. https://content.sierraclub.org/coal/
victories.

Siler, Wes. "Ryan Zinke Has Fired the DOI Inspector General." *Outside*, October
15, 2018. https://www.outsideonline.com/2355936/zinke-fires-inspector
-general.

Sissell, Kara. "House Leaders Release Draft Climate Legislation." *Chemical Week*,
no. 32 (2008): 10.

Smil, Vaclav. *Energies: An Illustrated Guide to the Biosphere and Civilization*. Cambridge, MA and London: MIT Press, 1999.

Smil, Vaclav. *Energy Transitions: History, Requirements, Prospects*. Santa Barbara:
Praeger, 2010.

Smith, Jordan Michael. "Northern Promises." *World Affairs* 176, no. 2 (July-
August 2013): 72–79.

Staub, Colin. "EPA's Recycling Funding Untouched in Final Budget." *Resource
Recycling*, March 27, 2018. https://resource-recycling.com/recycling/
2018/03/27/epas-recycling-funding-untouched-in-final-budget/.

Stevens, William K. "Meeting Reaches Accord to Reduce Greenhouse Gases." *New York Times*, December 11, 1997. https://www.nytimes.com/1997/12/11/world/meeting-reaches-accord-to-reduce-greenhouse-gases.html.

Stevens, William K. "Split Over Poorer Countries' Role Puts Cloud on Global-Warming Talks." *New York Times*, December 6, 1997. https://www.nytimes.com/1997/12/06/world/split-over-poorer-countries-role-puts-cloud-on-global-warming-talks.html.

Stocker, Thomas, Dahe Qin, Gian-Kasper Plattner, Melinda M. B. Tignor, Simon K. Allen, Judith Boschung, Alexander Nauels, Yu Xia, Vincent Bex, and Pauline M. Midgley, eds. *Climate Change 2013: The Physical Science Basis, Contribution of Working Group I to the Fifth Assessment Report of the Intergovernmental Panel on Climate Change.* Cambridge, UK: Cambridge University Press, 2013.

Stradling, David, and Peter Thorsheim. "The Smoke of Great Cities: British and American Efforts to Control Air Pollution, 1860–1914." *Environmental History* 4, no. 1 (January 1999): 6–31.

"Terrifying! Improbable! Chemtrails!" *Skeptic (Altadena, CA)* 22, no. 2 (Spring 2017): S64.

Tucker, Raymond R. *The Los Angeles Smog Report.* Los Angeles: Times-Mirror, 1947.

Union of International Associations. "Intergovernmental Negotiating Committee for a Framework Convention on Climate Change (INC/FCCC)." Open Yearbook. https://uia.org/s/or/en/1100025172.

United Nations Framework Convention on Climate Change. "About the Secretariat." https://unfccc.int/about-us/about-the-secretariat.

United Nations Framework Convention on Climate Change. *Paris Agreement.* https://treaties.un.org/Pages/ViewDetails.aspx?src=TREATY&mtdsg_no=XXVII-7-d&chapter=27&clang=_en.

United Nations Framework Convention on Climate Change. "What Are Bodies?" 2018. https://unfccc.int/process/bodies/the-big-picture/what-are-bodies.

United Nations Treaty Collection. "Chapter XXVII, Environment. 7.a. Kyoto Protocol to the United Nations Framework Convention on Climate Change." Status of Treaties. https://treaties.un.org/Pages/ViewDetails.aspx?src=TREATY&mtdsg_no=XXVII-7-a&chapter=27&clang=_en.

Urbina, Ian. "Piles of Dirty Secrets behind a Model 'Clean Coal' Project." *New York Times*, July 5, 2016. https://www.nytimes.com/2016/07/05/science/kemper-coal-mississippi.html.

U.S. Conference of Mayors. "Mayors Climate Protection Agreement." Mayors Climate Protection Center. https://www.usmayors.org/mayors-climate-protection-center/.

U.S. Conference of Mayors. *U.S. Mayors Report on a Decade of Global Climate Leadership: Selected Mayor Profiles.* Washington, DC: U.S. Conference of Mayors, December 2015. http://www.usmayors.org/wp-content/uploads/2017/06/1205-report-climateaction.pdf.

U.S. Department of Energy. Alternative Fuels Data Center. "Renewable Fuel Standard." https://www.afdc.energy.gov/laws/RFS.

U.S. Energy Information Administration. *Annual Coal Report 2015.* November 2016. https://www.eia.gov/coal/annual/.

U.S. Energy Information Administration. "Electric Power Monthly." https://www.eia.gov/electricity/monthly/epm_table_grapher.php?t=epmt_1_1_a.

U.S. Energy Information Administration. *U.S. Coal Reserves.* 2016. https://www.eia.gov/coal/reserves/.

U.S. Office of Personnel Management. "FedScope: Employment Cubes (September 2018)." Posted March 19, 2019. https://www.fedscope.opm.gov/employment.asp.

U.S.-China Joint Announcement on Climate Change. Press statement. Washington, DC: White House, Office of the Press Secretary, 2014. https://obamawhitehouse.archives.gov/the-press-office/2014/11/11/us-china-joint-announcement-climate-change.

Valentine, Mark. "The Twelve Days of UNCED." *Earth Island Journal* 7, no. 3 (Summer 1992): 38–40.

Vidal, John. "Plane Speaking." *Guardian*, November 1, 2006. https://www.theguardian.com/environment/2006/nov/01/travelsenvironmentalimpact.localgovernment.

Watkins, Eli, and Clare Foran. "EPA Chief Scott Pruitt's Long List of Controversies." *CNN Politics*, July 5, 2018. https://www.cnn.com/2018/04/06/politics/scott-pruitt-controversies-list/index.html.

Waxman, Henry A. "An Overview of the Clean Air Act Amendments of 1990." *Environmental Law* 21, no. 4 (1991): 1721–816.

Webster, P. J., G. J. Holland, J. A. Curry, and H.-R. Chang. "Changes in Tropical Cyclone Number, Duration, and Intensity in a Warming Environment." *Science* 309, no. 5742 (September 15, 2005): 1844–46.

Wheeler, Lydia. "Meet the Powerful Group behind Trump's Judicial Nominations." *The Hill*, November 16, 2017. https://thehill.com/regulation/court-battles/360598-meet-the-powerful-group-behind-trumps-judicial-nominations.

Wilkins, E. T. "Air Pollution and the London Fog of December, 1952." *Journal of the Royal Sanitary Institute* 74, no. 1 (January 1954): 1–21.

Wirth, Tim. "Interview." *Frontline*, April 24, 2007. https://www.pbs.org/wgbh/pages/frontline/hotpolitics/interviews/wirth.html.

Wolfson, Sam. "The Ethics Scandals That Eventually Forced Scott Pruitt to Resign." *Guardian*, July 5, 2018. https://www.theguardian.com/us-news/2018/jul/05/scott-pruitt-what-it-took-to-get-him-to-resign-from-his-epa-job.

Woloszyn, Andy, and David Finkel. "Rallying to Stop Keystone XL: Tipping Point for a Movement?" *Against the Current* 28, no. 1 (March-April 2013): 7.

Worland, Justin. "Donald Trump Does Not Believe in Man-Made Climate Change, Campaign Manager Says." *Time*, September 27, 2016. http://time.com/4509488/donald-trump-climate-change-hoax/.

World Commission on Dams. *Dams and Development: A New Framework for Decision-Making*. London: Earthscan, 2000.

World Commission on Environment and Development. *Our Common Future*. Oxford: Oxford University Press, 1987.

Zengerie, Jason. "How the Trump Administration Is Remaking the Courts." *New York Times*, August 22, 2018. https://www.nytimes.com/2018/08/22/magazine/trump-remaking-courts-judiciary.html.

Zieger, Robert H. *The CIO: 1935–1955*. Chapel Hill: University of North Carolina Press, 1995.

Zieger, Robert H. *John L. Lewis: Labor Leader*. Twentieth-Century American Biography Series, no. 8. Woodbridge, CT: Twayne, 1988.

Zinke, Ryan, and Scott McEwen. *American Commander: Serving a Country Worth Fighting for and Training the Brave Soldiers Who Lead the Way*. Nashville: Thomas Nelson, 2016.

Zola, Émile. *Germinal*. Les Rougon-Macquart. Paris: G. Charpentier, 1887.

Index

About the Author

John C. Berg is professor emeritus at Suffolk University, where he taught political science and environmental studies. Earlier in his career, he chaired the Government Department, and he directed the environmental studies program. His research has been increasingly focused on environmental issues in recent years, but he has also conducted research on legislative and party politics. Berg is author of *Unequal Struggle: Class, Gender, Race, and Power in the U.S. Congress* and editor of *Teamsters and Turtles? U.S. Progressive Political Movements in the 21st Century.*